StrateJEE

to help you prepare
well and win!

Dear eSaralite,

Welcome to the eSaral family!

You have taken a smart decision to prepare for India's highly competitive examination and can confidently count on us now to facilitate your preparation during this journey.

It is our mission at eSaral to make education Affordable, Accessible and Thoughtful. We provide personalised attention to every student and teach in a way that makes them visualise concepts, arouses curiosity, expands imagination and makes the student think critically. We make learning easy, fun and intuitive.

Our top-quality video lectures, comprehensive study material and tests created by Kota's experienced Top IITian and Doctor faculties alongwith 4 layered doubt solving and 3 layered personalised mentorship will help you in getting admission in your dream college.

The key to succeed in a challenging exam is to prepare well and prepare with the experts who have been there and who understand the intricacies of the exam, changing patterns and approach of the examiners. Our highly competent and experienced faculty members help you learn and master the strategy for preparing and cracking the exam. This book is the first step in that direction and we encourage you to read it, re-read it and keep referring to it to be on top of your StrateJEE game.

You will also be proud to know that eSaral family is growing at a rapid rate and our past students have already started bringing laurels to the family. We are excited to have you with us and it's a matter of great pride for us that you have chosen us to be your partners in success. We assure you that we will work tirelessly and will leave no stone unturned to ensure that you build a strong foundation which will help you achieve your future goals.

Welcome onboard, happy learning and wish you all the best!

Always with you,
Team eSaral
www.esaral.com

StrateJEE

Strategies used by
25000+ IITians & Doctors

First Edition (April)	2018
Second Edition (March)	2021
Third Edition (June)	**2023**

StrateJEE

Strategies used by 25000+ IITians & Doctors

Third Edition

Saransh Gupta
AIR-41, IIT-JEE (IIT Bombay, CS)

Prateek Gupta
B. Tech. IIT Bombay

arihant
SUCCESS AND NOTHING LESS

arihant
SUCCESS AND NOTHING LESS

ARIHANT PRAKASHAN (Series), MEERUT

卐 Administrative & Production Offices

Regd. Office
'Ramchhaya' 4577/15, Agarwal Road, Darya Ganj, New Delhi -110002
Tele: 011- 47630600, 43518550

卐 Head Office
Kalindi, TP Nagar, Meerut (UP) - 250002
Tel: 0121-7156203, 7156204

卐 Sales & Support Offices
Agra, Ahmedabad, Bengaluru, Bareilly, Chennai, Delhi, Guwahati, Hyderabad, Jaipur, Jhansi, Kolkata, Lucknow, Nagpur & Pune.

卐 PRICE ₹199.00

Published by Arihant Publications (India) Ltd.

For further information about the books published by Arihant, log on to www.arihantbooks.com or e-mail at info@arihantbooks.com

Follow us on

Preface

My two-year JEE preparation journey was full of ups and downs with ample of learning for me. I consider myself blessed to have guidance of my father and teachers who mentored and motivated me at every step. I also had access to the "Bhaiya Log" who had cracked JEE with great ranks and they were a source of inspiration for me. I used to visit them, talk about my preparation and discuss the problems that I faced. It was exciting to know about the tricks and strategies they shared with me. Learnings from their experience helped me avoid falling prey to the mistakes commonly made by aspirants.

During my JEE preparation phase I worked hard with a systematic approach. I developed strategies for time management, effective revision, test analysis and many such things, alongwith maintaining study material in an organized way. I am sure my JEE result would not have been as good without the strategic preparation.

My belief in need of strategic preparation and experienced mentorship alongwith hard work got strengthened during 8 years of teaching and mentoring thousands of JEE aspirants in Kota. Due to improper planning and lack of guidance, many students in spite of putting hours and hours of study are not able to get results which their hard work deserves. Based on my JEE journey and experiences of other JEE toppers, I worked on the strategies, taught
them to students in Kota, closely observed their impact, modified them and after iterations and testing came up with methodologies that resulted in better performance of students. The impact created by such strategies and their demand among students inspired me to share this knowledge in form of a book.

To get the most out of this book, I would recommend the aspirants to not just go through the strategies but to put them to practice in their preparation and see, if the strategies work out for them. I hope the strategies in this book will help you get the most out of your precious hard work and make your JEE journey more cheerful.

Cheers!

Acknowledgement

I would like to express my gratitude to each and every person who had contributed in making this book possible. First and foremost, I would like to thank my family for their constant support and encouragement in this endeavour, particularly my younger brother Prateek and my wife Shruti. Without them, this book wouldn't have been possible.

I am grateful to Bhavana ji for constant feedback, editing of the book and most importantly for her belief in the book.

A big thanks to Isha, Sanskriti and Shruti for creating such beautiful and innovative illustrations.

I am thankful to Achin, Sai, Palak, Amit, Arpit, Shashank, Amrose and Lav for sparing their valuable time and sharing their JEE journey.

I am indebted to my students from whom I have learned so much. Their eagerness to read the book was a constant push in timely completion of the book.

I would like to thank Kanika for helping me with the publishing process.

Sincere thanks to Arihant Publication team for giving the finishing touch to the book and publishing it.

About the Author

Saransh Gupta
Computer Science from IIT Bombay

Saransh Gupta is a graduate in Computer Science from IIT Bombay. He cracked IIT-JEE in 2006 with All India Rank (AIR)-41 along with securing AIR-71 in AIEEE (JEE Mains). He has also been awarded Mamraj Agarwal Award by Smt. Pratibha Patil. He has an experience of working with companies like Google, Oracle, Goldman Sachs alongwith having research experience at Max Planck Institute, Germany. Thereafter, to pursue his passion in teaching, he joined Allen Career Institute in his home town, Kota, as a Physics faculty to coach JEE aspirants. He has mentored thousands of students and under his guidance many students have fulfilled their dreams and got into premier institutes.

After 5 years of experience, he left his coveted job and started working for a noble mission. He has co-founded a startup eSaral (www.esaral.com) to provide equal educational opportunity for all by making education accessible, affordable and thoughtful.

Author can be contacted at saranshgupta@esaral.com or https://www.facebook.com/saransh.iitb.

About the Author

Prateek Gupta

B.Tech from IIT Bombay

Prateek Gupta Sir, graduated from IIT Bombay, is an esteemed educator and co-author of the renowned "StrateJEE" book. With his strong academic foundation, Prateek Sir has played a pivotal role in guiding students towards success in competitive examinations. What makes him truly remarkable is his decision to give up a placement job offer worth lakhs of rupees from IIT Bombay. Instead, he chose to dedicate his life to helping students in need. Prateek, his selfless commitment to education and the well-being of students is incredibly inspiring. By sacrificing personal financial gain, he made a conscious choice to focus on supporting and guiding students who need it the most. His decision shows his deep passion and unwavering dedication to making a positive difference.

Under his mentorship, students like Deepak, Antara, and Apurv have experienced extraordinary transformations. Deepak witnessed a remarkable leap in his JEE Main scores, progressing from the 45th percentile to an impressive 99th percentile. Antara, despite facing personal hardships including the loss of her father and illness, displayed remarkable resilience and achieved an astounding jump from the 15th percentile to an exceptional 99.95th percentile in the JEE Main. Similarly, Apurv showcased significant progress, elevating his scores from the 33rd percentile to an outstanding 99th percentile.

Prateek Gupta Sir's expertise, coupled with his unwavering dedication, has not only impacted the lives of individual students but has also paved the way for aspiring engineers and medical professionals in rural and tribal areas to realize their dreams. His contributions, celebrated within the pages of the esteemed "StrateJEE" book, stand as a testament to his commitment to educational excellence.

Author can be connected at prateek@esaral.com

Brief Contents

Contents

Contents

01

Introduction

It was 31 May 2006. The results of IIT-JEE (Indian Institute of Technology - Joint Entrance Exam) 2006 were to be announced. To check the JEE result, I went to a relative's house; they had a computer with a 256 Kbps internet connection (mobiles and laptops were not common back then).

Since it was the day of results, the traffic on the JEE website was heavy. I waited with bated breath, getting impatient for the page to load. It did and I entered my roll number. The result page started loading slowly. With each line showing up, my heart pumped more intensely. Finally, my All India Rank (AIR) came up: AIR-41.

I was born and brought up in Kota, Rajasthan. Yes, it is the city famous for IIT-JEE and pre-medical coaching as also for its *kachori* and *Kota saree*. Every year the city is home to more than a lakh candidates preparing for engineering and pre-medical entrance exams.

IIT-JEE is one of the most competitive entrance exams. The competition can be gauged by the fact that in the year 2017, over 12 lakh aspirants appeared for the JEE Main exam, from which 2.2 lakh candidates qualified for the JEE Advance exam. Out of these 2.2 lakh aspirants who appeared for the JEE Advance exam, a little over 10,000 finally got selected for the IITs.

The questions that arise here are: What do these 10,000 candidates do different from the rest and how do they do it? What is the 'secret' especially of top rankers?

To get closer to the answer, we have to first understand that preparation for IIT-JEE is a marathon run of two or more years involving mastering the huge syllabus of three subjects— physics, chemistry, and maths (PCM)—from Classes 11 and 12. In all, there are around 90 chapters in these three subjects. Sounds scary?! However, the silver lining is that *IIT-JEE can be cracked*. It is cracked each year by many candidates.

The basic requirements are hard work and perseverance. Without these, success is unthinkable. Further, to get the best out of the hard work involved, proper planning and strategies are essential to channelize your efforts and hard work in the right direction.

As the competition is fierce and JEE permits only two attempts (one in Class 12 and one year after class 12), strategic preparation is needed. Things cannot be left to chance. For instance, strategy is needed to systematically manage the huge study material of JEE, for effective revision, to maximize the test scores, to avoid silly mistakes, for the Class 12 Board exams, for the last month before JEE, etc.

Every strategy that can improve your result will matter. During my preparation for JEE-2006, these strategies played a very important role. Be it systematic markings in notes, assignments, etc. or test strategy, time management, or body training during the last month before JEE—everything counted in getting me AIR-41.

Till year 2005, the pattern of JEE was as follows: First qualifying round of objective questions (known as *Screening*) followed by a second round of subjective papers for PCM. So, we prepared according to this pattern for one-and-a-half years till, in the middle of Class 12, JEE announced a change in its pattern from 2006 onwards.

Screening was removed, and there was going to be only one round in JEE, that too OMR-based.multiple-choice questions only, where the answer sheet is based on optical mark recognition (OMR), which requires one to shade the dot indicating the correct answer.

This announcement initially led to panic, as our preparations and test practices were for a different pattern, but the relief was that this change was for all candidates! Pressure was mounting, as the Board exams were also approaching. As it is said, *"Musibat aati hai to har taraf se aati hai."* (When problems come, they come from all directions). I caught Jaundice. Weakness took over my whole body and sometimes the condition was so bad that I had to miss classes. I was not able to prepare at will as rest was required for recovery. It was a difficult time. Along with support from family and friends, systematic preparation came to my rescue. Organized work, marked notes, assignments, NCERT textbooks, etc. facilitated revision in a short time.

When I reflect on the preparation days, I realize that each of my friends who got ranks under 100 in JEE had some or the other strategy, which supplemented their hard work in getting marvellous results. And I must admit that many of the strategies were learned from each other during discussions. In the preparation pattern followed by different people, the common thing was: The preparation was not random. It was strategic and systematic. I have taught physics in one of the renowned coaching institutes in Kota, and have observed the same pattern during my years of teaching. Having mentored more than 5000 IIT-JEE aspirants, my belief in strategic preparation has strengthened day-by-day. Candidates who do strategic hard work generally perform better than those who do the same hard work but randomly.

I will share a situation that is very commonly observed among candidates. Consider three candidates A, B, and C. All of them solve each and every problem sincerely, but the difference lies in how they do it. A solves everything in rough papers. B solves them in a proper notebook, with a proper solution for each problem but does not mark any question indicating its level of importance, e.g., with a star or something similar. C also solves them in a proper notebook along with marking every question according to its difficulty level, say, easy, good, or very good.

All three did nearly the same hard work (C a bit more than B and A). Now, at the time of revision, when there are many chapters to revise, who can revise effectively? C will just pick up the 'very good' or 'good' marked questions and go through all the 'good' problems, going through the solution in case he/she gets stuck and finish his/her revision quickly. B has to go through each problem, as he/she does not have any markings. And what about A? He/she would have nothing to refer to; neither solutions nor markings.

So, despite the same initial hard work, C will have an upper hand due to *planned hard work*. Consider the effect of this strategy when you have to revise 90 chapters!! Now this seems to be an obvious thing, but believe me, majority of candidates fall in the A and B categories, not because they do not want to do systematic work, but because they do not realize the importance of it, and in the first place they are not aware of such strategies.

The path to the IITs is mostly similar for all aspirants if we talk in terms of syllabus, pattern of the paper, etc. and these have been the same for many years. I would like to share that still more than 30% of the questions, that students practise, are the same as they were back in 2006. When each year, lakhs of candidates are going through the same path, some failing, some succeeding, and some outstanding, would it not be wise to be aware of the common pitfalls and pick up the strategies that led people to success.

Before entering into the field of teaching, I have worked at Google, Oracle, and Goldman Sachs. The corporate world experience was very enriching. It taught me the importance of strategy, systematic work, and planning to make any project a success.

And preparation for JEE is no less than a project!! During my teaching sessions with IIT-JEE aspirants, I reserved 30 minutes at the end of Saturday's class once in two weeks to discuss the strategies and methodologies for systematic preparation. These methodologies have helped a lot of candidates gain the most out of their hard work and crack IIT-JEE with flying colours. Also, during the preparation phase, these strategies have led candidates at all levels to achieve incremental improvement from their current level of performance.

As the path of preparation is similar, so are the problems (most of them) faced by the candidates. To add to my experience of problems faced by candidates, I have conducted a written survey of 1000 candidates on the challenges faced by them. After classifying the challenges into major heads, I have conducted seminars on improve-ment strategies to address these problems. These seminars received very encouraging responses, and candidates showed marked improvements in their performances. This gave me the motivation to compile the experience-tested strategies and techniques addressing the challenges faced by candidates into a book.

The journey of JEE also has many ups and downs, as is the case with almost every rewarding journey. There is a pressure to perform and fear of failure, along with so many distractions. There are difficult times when things do not go as we would like, and they do come up in almost every aspirant's journey.

A few chapters are added in the book to help you sail through these difficult times, cope up with stress, and stay motivated. To gain the inputs from the experiences of people who cracked the JEE with high scores, the journeys of four people (including AIR-1 of JEE-2007 and JEE-2013) are included in the book. I am thankful to them for sparing time from their busy schedules and sharing their JEE journey with us. To get the most out of this book, I would recommend working out with the strategies mentioned in the book at least a few times. For your reference, here is an overview of the book:

- **Chapter 2** *Getting Ready for the JEErney* Well-begun is half-done! This chapter is about getting equipped for the adventurous journey of JEE. Changes/challenges encountered upon transit from the world of Board/school exams to the competitive JEE, subject selection in Class 11, and tips to get a head start are elements of this chapter.
- **Chapter 3** *Index the Path* The JEE syllabus combines that of both Classes 11 and 12. To proficiently manage the huge volume of study material and syllabus, the techniques of indexing and marking are discussed in this chapter.
- **Chapter 4** *The How of Revision* How to revise efficiently? How to complete pending work? How to prioritize among revision, homework, and backlogs? How to keep up with Class 11 topics along with Class 12? Looking for answers? This chapter may help you to find them.
- **Chapter 5** *Test Strategy* Attempting an exam paper is an art that is equally important as the subject knowledge. This chapter lays out various test strategies for improving performance and getting closer to the target you know you deserve.
- **Chapter 6** *Going Online* This chapter has some dos and don'ts and strategies for better performance while appearing for an online exam.
- **Chapter 7** *You Can Do Better* Analysing after a test is very important to get most out the test. The chapter describes the systematic approach to test analysis and how to utilize it for improvement in future tests. This chapter also details the methodologies to avoid silly mistakes and calculation mistakes, get rid of exam phobia, and improve speed along with accuracy.

- **Chapter 8** *Subject-wise Tips* JEE has three subjects (PCM), and tips to strike a balance among the three as well as tips for mastering each subject are dealt with in this chapter.
- **Chapter 9** *The Path to High Performance* This chapter is a guide to the path that leads to your goal by showing how you can improve concentration, get rid of distractions, cope up with stress, cultivate good habits, and utilize peer learning for high performance.
- **Chapter 10** *Time Management and Planning* If you are looking for effective time management, increasing self-study time as well as its productivity, setting up goals/targets, and most importantly keeping up with them, this chapter has innovative strategies for you to plan proficiently.
- **Chapter 11** *How Not to Give Up* Great journeys are rarely free of difficult times. What makes these journeys great is not giving up in the difficult times. This chapter will help you sail through difficult times and overcome fear, and will act as a source of inspiration to continue on your path towards success.
- **Chapter 12** *Those Who Triumphed* This chapter narrates the journey of hard work, perseverance, strategies, and methodologies employed by those who have successfully cracked the JEE (includes stories of AIR-1 of IIT-JEE 2007 and 2013).
- **Chapter 13** *The Last Month* A month's time between the Board exams and JEE has a very crucial role to play. During this time, how to review the complete two-year syllabus, condition yourself for the Big day, train the body as per the exam schedule, and tips for the day before the JEE and the day of the exam are detailed in this chapter.

- **Chapter 14** *Art of Writing Boards with JEE* This chapter presents a time-effective methodological preparation strategy for the Board exams along with preparing for the JEE at the same time. Guidelines on when and how to read NCERT books, writing the Board exam papers, and some more Board-specific tips are also included.
- **Chapter 15** *Life@IIT* A glimpse of life at the IITs, encompassing numerous opportunities and all-round development of personality, is presented in this chapter.

Let's get Started! My best wishes for your Journey. All the very best!

Scan the QR code to listen to the author on this topic

02

Getting Ready for the JEErney

On a fine Sunday morning, a friend of my father's came to visit us with his son Amit, who had just finished his Class 10 Board exams. Amit had decided to prepare for IIT-JEE. After some general chit chat, Amit's father asked me *"Amit ko JEE ki tayyari ke kuch tips de do"* (Do give Amit some tips for JEE Preparation.). Amit looked at me with excitement and high expectations like Kung Fu Panda ready for the Dragon Scroll!!!

He, however, became a bit puzzled, when I asked "Why JEE?" He said "I want to become an IITian". Then I asked "Why do you want to become an IITian?" He looked at his father expecting an answer and his father signalled him to speak for himself. He shyly answered "IITians get high pay packages." I had laughed within thinking of the media who showcase a pretty picture of the IITs. He was astonished when I said "If this is the only reason then there is no need to go to an IIT. You will earn much more in your father's business in a month than an IITian's average annual package that is offered during placements. Join your father business now!"

This set him thinking. After some introspection, he said "I will have a good reputation in the society, overall development, learning opportunity…" Then to know more about the IITs, he asked me about life at IIT, career opportunities after IIT, etc. Once he got convincing reasons for his decision, we discussed how to start preparing for JEE, the difference from school exams, and a few initial tips for about an hour. Then he left with a smile on his face, confident of his decision.

Why JEE?

Before starting the JEE journey, we should know why we are choosing this path and what we will get. The answer is not *just the IITs but what from the IITs?* The reasons should be clear and strongly convincing. These reasons will be our motivation that will make us walk through difficult times and excel in the journey. This path should be our own choice and not dumped upon us by others. We can take guidance and find out about the IITs from the Internet, friends, family, teachers, and relatives, but finally the choice and reasons should be ours.

The Journey

Once we have the conviction for choosing the JEE, the journey is going to be thrilling and a great learning experience. JEE preparation is different from school/Board exams preparation in the following aspects:

1. School exams have one year's syllabus while the JEE has the syllabi of two years (Classes 11 and 12).
2. The JEE is a low-scoring exam, where you may get selected with just over 40% marks, whereas in school or board exams, sometimes 90% marks are considered to be average. If you get 90% marks in the JEE, then you are most likely to end up with a top 10 All India Rank (AIR). Understand this clearly: *What matters in the JEE is the rank, not the percentage.*
3. Board exams are subjective and have a traditional pattern. The JEE is OMR-based (optical mark recognition) with innovative patterns that keep on changing. So, the preparation for the two exams is not exactly the same. The questions of the JEE are also tougher than those of the Boards.

One more important difference is in the *level of competition.* In school exams, competition is among a few 100 students or less, whereas in the JEE, about a million students compete. The good news is that there are more than 10,000 seats in the IITs.

As the competition is tough, training for the JEE will be more rigorous than that for Board exams. And sweeter will be the reward. The reward is not just getting admission into an IIT or NIT.

This is only one part. The other part that goes unnoticed is the learnings during the journey. The rigorous training that you will go through will nurture important qualities in you like perseverance, determination, hard work, time management, and many more, not to forget the development of a scientific mind, ability to analyse, and think outside the box. These precious qualities will be your assets in your future endeavours.

If you are preparing for the exams with a coaching class for the first time, and the class has hundreds of students, then the environment might affect you. In school, you might normally be in the top 10 among 60–70 students, but in coaching classes you would be competing on a much larger scale with toppers of other schools also. It will be a transition from local to national. In such a competition, you might not be in the top 10 like you were in school.

Don't lose heart. You will improve and the IITs have more than 10 seats!

The positive part of this competition is that you have smart people around you to learn from. This is an opportunity to learn and grow through working with highly intelligent peers. Take the competition as a thrust for you to perform better. Second, tougher the competition during the preparation phase, closer it is to the real scenario, i.e., in the JEE. *Competition is inevitable. What matters is how we take it.*

If you have moved away from your home for JEE preparation, sacrificing your comforts at home and leaving old school friends behind, know that it is a brave step towards your goal. Sometimes you might feel home sick, especially if it is the first time you are away from home. In such times, recall that you are on a mission, and nothing can come between you and your goal.

You have to be conscious about your health. Sickness will mean loss of time and studies. Eat healthy (less junk food!) and timely, drink a lot of water, and do not skip meals, especially breakfast. A little exercise/yoga and meditation helps a lot in being energetic and focused throughout the day.

The journey of JEE will be in no way ordinary, as ordinary paths do not lead to extra-ordinary results. The path will have turns, difficulties, and what not, but your reasons and motivation will not let you stop. *Remember, a diamond is not made in a day. It takes time and rigorous work to manifest. So will you.*

Getting Started

The first question that arises is: When to start preparing for the JEE? After Class 12 or from Class 11 or from Class 9 or even earlier?! As the IIT-JEE window is open only for two years for a candidate, it is better to start from Class 11, which most people do (though there are people who crack the JEE with an intense preparation of just one year after Class 12).

Starting from Class 9 can give one a head start, but one should take the precaution of not overloading Classes 9 and 10 with the syllabi of Classes 11 and 12. The focus in these junior classes should be on building concepts rather than just cramming the formulae of higher classes without understanding them. The habit of cramming without understanding concepts may give short-term gains in some exams but is not of much use in the JEE.

If you decide to start preparing for the JEE after Class 10, then you will have Physics, Chemistry and Maths (PCM) as three of your subjects in Class 11. Most of the Boards, including CBSE, have the fourth subject as English. The fifth subject is an elective (some Boards may have more than 5 subjects). The same five subjects continue in Class 12. The fifth subject should be chosen keeping in mind that most of your time will be spent in PCM. The fifth subject should be the easiest one, which can be prepared for in the least time. I took physical education (PE) as the fifth subject (CBSE has this subject, but some of the Boards do not have PE as an option). PE saved a lot of time for me during JEE preparation. Most of the JEE aspirants under CBSE choose PE as the fifth subject.

Self-study plays a crucial role in preparation. For effective self study, you should have a favourable place to study. If it is possible, fix a room dedicated for your studies with the least disturbance. Set your room before you start your JEE preparation. Put up some motivational posters or quotes in your study room.

In your preparation phase, as you would be learning new concepts, many academic doubts will arise for sure. These doubts are very important. Do not ignore them and, most importantly, do not hesitate to ask and clarify them. Doubts by no means imply that you are dumb, they mean that you are concerned about your studies, you are thinking, you want to know more. A doubt, if not resolved at the appropriate time, will create problems in clear understanding of concepts in future. And these doubts have a cunning trait, they show their faces in the exam! Do not feel shy to clarify your doubts. You are there for knowledge and learning. Doubts will definitely come. They can come in the classroom, during self-study and discussions, and many a times in the washroom. Keep a pocket diary with you (obviously outside the washroom!). Write your unresolved doubts in it. Carry this diary with you to the classroom. Sometimes, if during a class, you are not able to get an opportunity to clarify your doubts, note them down in your diary. Get them resolved whenever you get an opportunity. But beware that your doubts should not become a garb for spoon feeding. Work on the issue yourself before asking your teacher or friends. This will build your inner strength and confidence.

There will be another function of your diary: to note homework and *Think Upon* problems. Along with giving the routine homework, many a times your teacher will ask you to think upon some concept/problem (which I refer to as *Think-Upon* problems) for the next class. These problems will help you to better understand the concepts. To remember spending time on these *Think Upon* problems after class, note them in your diary during your class time. Do not ignore them.

The small seats-to-applicants ratio makes JEE highly competitive. The journey of JEE is, however, adventurous. Every arrow of tips and strategy in your quiver will take you towards piercing your target. Hard work has no substitute, but it has to be smart. As the time for preparation is limited, planning plays a vital role. Failing to plan would mean planning to fail. Become a smart test warrior.

The upcoming chapters will map out strategies, tips, and systematic approaches, along with the methodologies adopted by JEE toppers to help you navigate the waters and reach your destination joyfully. These chapters will also help you triumph over challenges that you may face along your JEE journey, and perform better, better, and better.

Let the Journey begin…

Scan the QR code to listen to the author on this topic

03

Index the Path

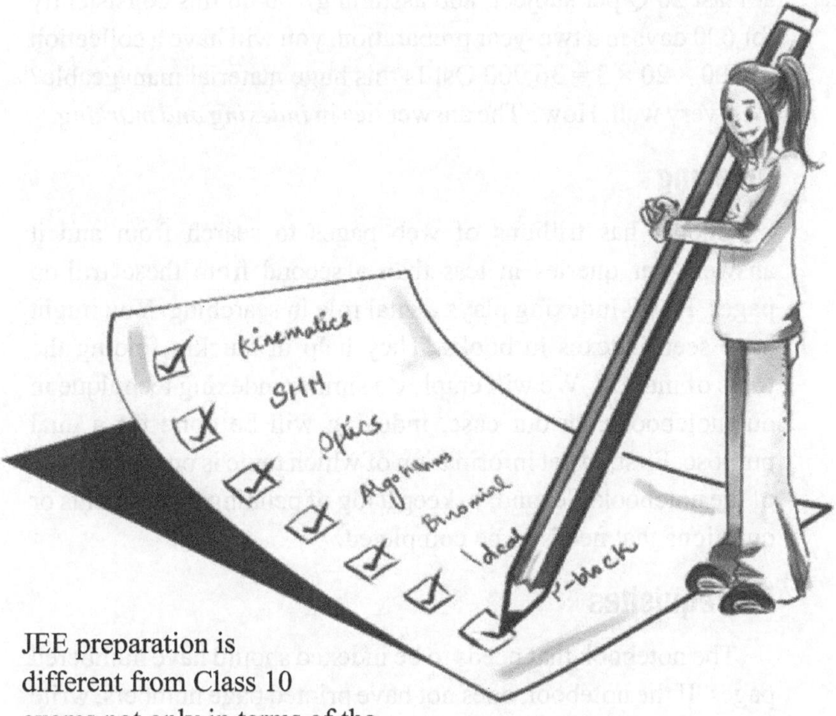

JEE preparation is different from Class 10 exams not only in terms of the pattern, level of questions, and competition but also in terms of the bulk of syllabus.

In Class 10, there are around 15 chapters in Maths and 5 each in Chemistry and physics. On the other hand, JEE preparation means dealing with around 4000 pages of notes and 36,000 questions in physics, chemistry and maths (PCM)!!! Surprised? Let us see.

The JEE syllabus combines that of both Classes 11 and 12. Each class has around 15 chapters in each of the subjects PCM. That means 15 × 3 × 2 (both classes) = 90 chapters. Each chapter would have an average of 40–45 pages of notes implying around 4000 pages.

And let us not forget the questions (Q)! If each day you solve at least 20 Q per subject, and assuming you do this consistently for 600 days in a two-year preparation, you will have a collection of 600 × 20 × 3 = 36,000 Qs! Is this huge material manageable? Yes, very well. How? The answer lies in *indexing and marking.*

Indexing

Google has trillions of web pages to search from and it answers our queries in less than a second from these trillion pages. How? Indexing plays a vital role in searching. You might have seen indexes in books. They help in quickly finding the topic of interest. We will employ a similar indexing technique in our notebooks. In our case, indexing will be done for a dual purpose. First, to get information of which topic is on which page of the notebook. Second, to keep a log of pending assignments or questions that needs to be completed.

Prerequisites

The notebook that needs to be indexed should have numbered pages. If the notebook does not have printed page numbers, write them manually but in a smart way. Do not number 1, 2, 3, 4, 5, 6…, instead number only the odd numbered pages 1, 3, 5, 7, 9… This will reduce your efforts by half.

Indexing Notes Copies

Make an index at the beginning of the notes copy. The index will have the topic names, subtopic names, and the respective starting page numbers of these topics. A sample index is shown in Table 3.1. When you will have a pile of notebooks to search from and you want to revise the notes of one subtopic, the index will guide you to jump directly to the relevant pages that contain the subtopic.

This will save time and effort during revision. This will also reduce stress and avoid panic arising from not being able to find the topic you need to revise.

TABLE 3.1 Sample index

S No.	Topic	Pg No.
⋮		⋮
3	Newton's Laws of Motion	65
3.1	Forces in Nature	65
3.2	Three Laws of Motion	67
3.3	Free Body Diagram	72
⋮		⋮
4	Friction	113
4.1	Static Friction	114
4.2	Kinetic Friction	120

Indexing Assignment Copies

Along with the theory, questions have to be solved for JEE preparation. Depending on whether you are preparing with some coaching centre, or through online videos, or self-study through books, or a combination of these, you will have collection of questions for practice in different formats by different names. These collections or assignments can broadly be segregated into three categories:

1 **Sheet / module / question bank for a topic** These are topicwise modules that have a collection of questions on a topic. There could be many exercises in them either based on the level or pattern of questions. A sample index for such modules is shown in Table 3.2.

TABLE 3.2 Sample index for topic-wise modules

S.No.	Topic Exercise	Pg. No.	Status
3	Newton's Laws of Motion	45-90	
	Exercise-I	45	Complete
	Exercise-II	63	Pending - Q 3, 7, 10
	Exercise-III	78	Pending - Q 5, 13

The Status column will indicate the status of completion of an assignment. The Status will be 'Completed' when an exercise is completed and 'Pending' when we could not complete some of the questions that were given in homework due to any reason. The status will keep a log of the pending work for us and will also act as a reminder! Write 'Pending' Status with a pencil, so that when we complete the work, the Status can be changed to 'Completed'.

2 **Daily practice problems (DPP)** These are assignments having a small set of questions (about 10-15 questions) that is given each day for practice. It may have questions from more than one topic. A sample index is shown in Table 3.3.

TABLE 3.3 Sample index for daily practice problems

DPP	Topic Exercise	Pg. No.	Status
1	Unit & Dimension 6 (U & D)	3	Complete
2	Unit & Dimension	6	Complete
3	Kinematics, U & D	8	Pending - Q 3
⋮	⋮	⋮	⋮
13	Newton Laws, Kinematics	38	Complete
⋮	⋮	⋮	⋮

The column 'Topics' will have the name of topics from which there are questions in that DPP. When we will have, say, more than 50 DPPs with us and we want to revise questions from the chapter Kinematics, we do not have to search for questions in each DPP. The index will save a lot of time. We will open the index and see which all DPPs had questions from Kinematics and revise the relevant questions from the respective DPPs. Done!

3 **Books** Some books (NCERTs, books by authors like H.C. Verma, I.E. Irodov, Hall & Knight, and many more) are used as reference books for the JEE. For questions from these books, make an index like that for topic-wise modules (Table 3.2).

For any other format, you can make a customised index on similar lines, having the Pg No. and Status columns.

An Important Note In my years of teaching, I have found many candidates solving questions in rough notebooks/pages. As rough pages do not have questions in any particular order, indexing, marking, and track-back is difficult. Such rough work cannot be reused during revision. Allot a notebook for modules, a notebook for DPPs, and one for books. Solve the questions in respective notebooks at appropriate locations in neat handwriting, and maintain the index. This will make your work systematic and, thus, reusable during revision.

As the syllabus for the JEE is large, the concept of last-minute preparation does not apply for the JEE. It is year-long systematic work that matters. During the time of revision, you should easily find what you have solved before. For this, the index will come into play. Fill the index regularly in small chunks whenever you find time. This way the index will be filled before you realize! You may use some other mechanism depending on your suitability and likings. For instance, if you like to work with loose pages instead of notebooks, go ahead, but take care not to lose them.

Feel free to apply your own creativity to create a structure/mechanism that works for you. Any method that will lead you to easily find what you have solved before will do. It is worth devoting time to make your academic work organised and reusable.

Marking

Imagine yourself in the duration of around 45–50 days between the JEE Mains and JEE Advance exams. You have a stack of around 4000 pages of notes and 36,000 questions to revise. Is it possible to go through every word of these notes and re-solve every question in this material in this limited duration? No. The more important question: Is everything *worth* revising? The answer would most probably be No. So, what to revise? Only the topics and issues important to you. But how to identify the important content without going through everything? The answer lies in *marking*. Marking is to be done in both assignments and notes.

Marking in Assignments

Some of the questions in the assignments are basic formula based and are easily solved. Their purpose is to help you understand the basics of a chapter and learn the application of formulae. They are needed at the beginner's level for you to get a hold on the chapter. Once you get a hold of the chapter, you do not need revision of such questions. Mark them as *Easy (E)*.

Then there are questions which take time to solve and have important concepts involved. You learn considerable things from them. You might be able to solve them in the first attempt. Mark such questions as *Good (G)*. Then there are the 'Wow' questions, which have multiple concepts involved or have some tricks involved or are unsolvable by you. You definitely want to revisit them. Mark such questions as *Very Good (Vg)*. So, we have three categories: Easy (E), Good (G), and Very Good (Vg).

Mark the questions as soon as you attempt them. Marking of questions may have to be updated when you revisit the questions later on. For example, question that seemed *Good (G)* at the first attempt may have now become trivial after solving the same concept many times. So, update its marking from G to E. A sample is shown in Fig. 3.1.

E **Q1.** A player caught a cricket ball of mass 150 g moving at a rate of 20 m/s. If the catching process is completed in 0.1 s., the force of the blow exerted by the ball on the hand of the player is equal to

(A) 150 N (B) 3 N (C) 30 N (D) 300 N

g E **Q2.** Three blocks of identical masses are connected by strings as shown. The whole system is accelerated to the right on a frictionless, horizontal surface by a horizontal surface by a horizontal force F. The net force acting on the middle block is

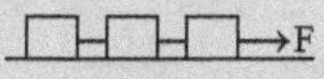

(A) zero (B) F (C) 2F/3 (D) F/3

g **Q3.** The wedge of mass M sits at rest on a scale as shown in figure. A small block of mass m slides down the frictionless incline of the wedge.The scale reading while the block slides, is

(A) $(m+M)g$ (B) $Mg + mg\cos^2 q$
(C) $Mg + mg\sin^2 q$ (D) $(M-m)\,g + mg\sin^2 q$

Vg **Q4.** Block A of mass m/2 is connected to one end of light rope which passes over a pulley as shown in the Fig. Man of mass m climbs the other end of rope with a relative acceleration of g/6 with respect to rope find acceleration of block A and tension in the rope.

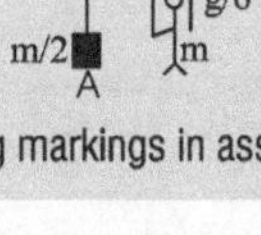

Figure 3.1 Updating markings in assignment questions

This is one way of marking. You can as well adopt some other ways of marking, such as star-based marking, where you assign stars to questions according to their level.

Marking in Notes

Notes are made up of theory, formulae, illustrations, definitions, and key points (important points). When notes are picked up for revision, does every section require equal attention? No. Some of the sections become trivial as we get a hold on the topic. These sections can be skipped when quick revision has to be done. But how to identify them without going through them? Mark them.

Mark sections that have become trivial for you as Easy (E) as shown in Fig. 3.2. Similarly, star-mark important sections that you find worthy of revision. These markings will keep you focused on the important sections and save time and brain processing in going through non-essential things. When I had

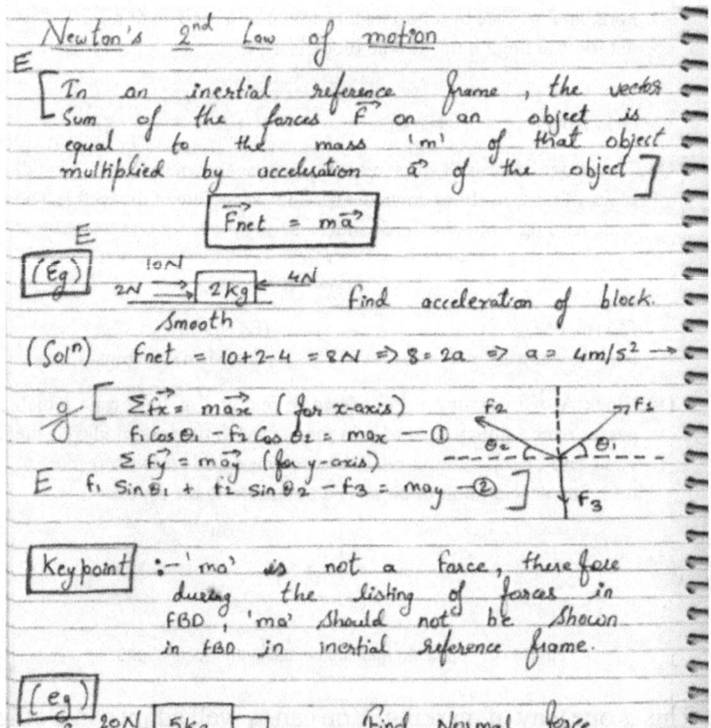

Figure 3.2 Marking in notes

revised my notes many times, some of the pages in the notes had become too trivial to revise. There was nothing in them worth revising. I used to mark a cross (X) on top of such pages. In the year 2006, when we had less than two weeks between the Boards and the JEE (which is now the JEE Advance) exam, these E, star, and cross markings were a saviour for me. Along with revision of all the important sections of notes, I could spare some time for practising questions in those two weeks.

For identifying formulae and key points quickly, highlight them in a *box*. When you need to revise the formulae and key points of a chapter, just follow the boxes and you will be through. These boxes will be like *Short Notes* for you (if making separate *Short Notes* will give you confidence and help in remembering and better understanding a topic, then go for them along with these markings). Mark the illustrations (Fig 3.2) in the notes as you would do for questions (E, G, Vg). In each revision, update the markings for the theory as well as the questions. If you follow some books along with your notes, mark them as well similarly.

Marking will prove to be a boon during revision. Whether it is revision for a practice test or the last-month revision for JEE when the entire syllabus has to be covered, these markings will help you revise as per the time available. When ample time is available, revise everything. However, when you want a quick revision of some topic, go through the highlighted formulae and key points, read the marked notes, and pick only the Vg-marked questions of that topic. Depending on the availability of time, you can pick the E-and G-marked questions along with the Vg-marked questions. Strategies for efficient revision are discussed in detail in the next chapter.

Scan the QR code to listen to the author on this topic

04

The How of Revision

Students preparing for a competitive examination often have the following questions:

- How to revise?
- How frequently to revise?
- What all to revise especially a day before the test?
- How to deal with weak topics?
- How to complete pending work?
- How to manage the topics of Class 11 while studying in Class 12?

This chapter will address these questions.

Revision Strategy

When we revise for a test, we pick a topic and start revising using the notes we have taken. It so happens that as the test date arrives, we may have completed revising the notes of some chapters, but do not have time for practising questions. Generally, in every periodic test, this pattern repeats.

We use notes as the starting point because these are our comfort zone, and end up revising the notes of only those topics that we are already strong at. Is it worth the time spent? Does this process contribute significantly to our revision? Also, what strategy should one adopt for smart revision when there are many topics to be revised?

The WMS Technique

Do not just pick up a topic randomly and start revising. Revision is effective when done with some planning. The WMS technique is very useful. WMS stands for 'weak medium strong'.

To start with, classify the chapters of all the subjects as weak, medium, and strong. We will call a topic 'weak' if you are able to solve less than 30% of the problems in that topic area, 'medium' if you are able to solve 30% to 70%, and 'strong' if you are able to solve more than 70%.

For example, look at Table 4.1 below:

TABLE 4.1 Classifying topics as weak, medium, and strong

Kinematics	Strong	Logarithms	Medium
Newton Laws of Motion	Medium	Trigonometry	Strong
Friction	Strong	Quadratic Equations	Weak
Circular Motion	Weak	Sequences and Series	Strong
Work, Power, Energy	Medium	Binomial Theorem	Weak

After classification, start with a weak topic. Yes, a weak topic! Generally, weak topics are not among our favourites and we hesitate to pick them up for revision. That's why they remain weak! Whereas revising our beloved topics, i.e., strong topics, again and again is satisfying. However, as weak topics are also part of the syllabus (which is the sad part!), we would like to be in good terms with them and that too as early as possible.

So take a deep breath, gather some strength, and pick a weak topic. Start with your notes, read each and every line, and then practice the Qs that you have marked as Easy (as mentioned in Chapter 3). Now that you have become familiar with the topic, practise the G (Good) and Vg (Very Good) marked Qs. You may refer to your notes in between if you forget some formulae.

After the weak topic, pick a medium topic. Start with the notes but read only the marked sections of the notes to save time because you are already comfortable with this topic. Then practise directly the G- and Vg-marked Qs. If you feel like, you may practise a few Easy Qs also.

Once a medium topic is over, treat yourself with a strong topic. You do not need to read their notes, directly jump to the Vg-marked Qs along with some G-marked Qs. Repeat this cycle and take up Physics, Chemistry, and Maths in turn. Update the status of the topics after revision. For example, once you convert a weak topic into a medium or strong topic, change its status in your list. When you are revising, ensure that you update the markings in your notes and Qs as explained in Chapter 3.

Strategy for Revising Questions

When we revise a question, we generally solve it till we get the correct answer each time we do it. The time spent in solving involves thinking through the approach/concepts, writing equations, and doing the calculations. Now let us analyse this process minutely: For what purpose did we revise this question: learning concepts, equations, or calculations?

The answer would generally be "for the concepts and equations involved." The calculations are needed to verify whether or not the concepts/equations lead to the correct answer. But wait a minute! We have already verified this the first time around when we solved that question, and have it fully worked out in our solution notebook.

When ample time is available for revision, nothing is better than to again solve all questions till we arrive at the answers. However, when we have limited time, a smart strategy would be to take advantage of the solutions that we have already worked out. In this strategy, out of, say, 10 questions that we have picked for revision, we will solve only some of them (say, 5 or more, depending on the availability of time) fully till we reach the answer. For the rest, read the question, apply the concept, write the equations (without looking at the solution), and stop. Open the solution notebook and verify whether you have applied the concepts and equations correctly. We have thus gained most out of the question as well as saved the time involved in calculation.

Caution Solve at least some of the questions till the final answer for calculation practice.

Quick Revision on the Day Before the Test

On the day before the test, it usually feels as though we do not remember anything! Do not worry, it is all in the mind! For a quick revision, look only through the boxes marked in your notes (or short notes) having the formulae and key points. Keep this reserved for the time before you go to bed. After you sleep, the revision will continue in your dreams!

Homework or Revision or Backlog

As time is limited and we always find ourselves short of it, how do we invest it wisely—in doing the current-topics homework, revision, or in completing the backlog work? Current homework should be the top priority, and if it is completed on time, there will be no backlog.

However, things do not always go as we would like! Sometimes, due to unavoidable circumstances, our homework remains incomplete. The best way to handle this is to *make a rule* of not letting anything remain incomplete for more than 7 days. If you have no classes, on, say, Sunday, then reserve some time for clearing the backlog on that day. This will also act as a revision for you, and *you have now made your backlog a tool for revision!* Do not let anything incomplete cross a Sunday. So, week after week, you will be up to the mark.

Revision of Class 11 Syllabus While Studying in Class 12

In Class 12, a majority of the time is spent in preparing for Class 12 topics. However, as the JEE has syllabi from both Classes 11 and 12, we have to be familiar and at ease with Class 11 topics as well. Your worked-out solutions, indexes, and markings of Class 11 syllabus practice and notes will be your biggest assets. This systematic work will help you to revise more material in less time. For example, when you have to revise the notes of a chapter, revise only the marked notes. Similarly, for practising questions, depending on whether a topic is weak, medium or strong, revise the marked questions only.

The portion of time that was invested in keeping the work systematic has high RoI (Return on Investment).

Now that you know the 'how of revision' and I am sure you will do it well, let us find out how you can score up to your full potential, which your solid and smart revision truly deserves. Let us now work out the strategies for taking a test.

Scan the QR code to listen to the author on this topic

05

Test Strategies

Do you have a strategy for the JEE in periodic tests? Do you feel that you are getting lesser marks than you deserve? Do some questions that you could not solve during the test easily come to you outside the exam hall? Do you end up leaving some easy questions unseen due to shortage of time during the tests?

If these problems are bothering you, then this chapter is for you.

Handling the JEE paper is an art, where strategy plays an equally important role as knowledge. Proper test strategy will lead to maximizing your marks and getting closer to the target you know you deserve.

First of all, please know this: the JEE is a low-scoring exam. What does that mean? As per past year data, as shown in Table 5.1, scoring around 35–40% marks (in the General category) in IIT JEE has ended up in selection in the IITs. This implies that JEE is not as high scoring as the Boards.

If you get 50% marks in the JEE, you could end up in the top 1000 depending on the paper, but imagine what your reputation will be in your town if you get only 50% in the Board exams! That is the difference.

Strategy 1 Understand the difference and prepare yourself psychologically.

TABLE 5.1 All India Rank and the corresponding percentages scored

All India Rank (General category)	Percentage marks scored (roundedoff)		
	2014	2015	2016
1	93%	93%	86%
101	77%	72%	66%
501	68%	61%	54%
1001	63%	55%	49%
2001	57%	50%	44%
5001	49%	41%	36%
7001	46%	37%	33%
10001	42%	35%	30%

Strategy 2 In the JEE, 3 hours are given for each paper. If the time is increased (say, to 5 hours) for the same paper, many candidates can solve it fully. But due to the time limit of 3 hours, scoring 100% is next to impossible. In such an exam, the aim should be to *maximize* the marks rather than seek full marks. If in the process of maximizing your marks, you score 100%, then congratulations… you are AIR-1.

Strategy 3 If you have an easy question and a tough question, both carrying the same marks, which one would you solve first? Obviously, the easy question, because nothing extra is offered to you for solving the tough one. This is the basis of the strategy *50-30-50*. This strategy has benefitted many of my students.

What is the 50-30-50 strategy? In this strategy, you tackle the paper in two rounds. In the first round, 50 minutes are given each to physics (P), 30 minutes to chemistry (C), and 50 minutes to maths (M). In this round, we solve the easy Qs. If a Q seems very difficult or lengthy, it should be left in the first round and marked for attempt in the second round. Do not forget to mark the Qs. You can mark the unattempted Qs with a circle and the attempted ones with a tick.

In the first round, we will stick to the timelines allotted. For example, if 50 minutes of physics are over, we will strictly move to the next subject. This way, we ensure that the subject-wise cutoffs are cleared. Then, once we have attempted all the easy Qs in 50 + 30 + 50 = 130 minutes, then in the second round of approximately 50 minutes (total given time: 180 minutes – 130 minutes of first round = 50 minutes for second round), we crack the unanswered questions peacefully.

This strategy ensures that no easy Q is left unattempted due to lack of time and all easy marks are grabbed. In the first 130 minutes, we try to ensure a seat in an IIT, and in the next 50 minutes, with each Q solved, we build our rank.

I have experienced two other advantages with this strategy:

1 When a reasonable number of questions are already solved in the first round, our confidence gets a boost. This state is favourable during the second round for solving the tougher questions left out in the first round.

2 Sometimes, in the meantime (as you move on to the next question), the subconscious mind comes up with a solution that the conscious mind could not think of in the first round. You might also have experienced it—you slept with something unsolved and when you wake up, the solution is inside your head from somewhere! This is the magic of the subconscious mind.

Order of attempt of PCM

Strategy 4 This strategy is regarding the order of attempt of PCM. Different candidates have different attempt orders. The best order for you is the one that works for you. Till you have found *the best order*, experiment with different strategies to figure out what works for you.

Following are the recommendations that I found work for most of the candidates.

Among the three subjects PCM, physics and maths require problem solving but in chemistry, most of the inorganic and organic portion is knowledge-based. Either you know them or you do not. So ideally, chemistry should not be given more than 45 minutes in all (unless the questions are exceptionally tough). I do not recommend starting the paper with chemistry for two reasons.

First, chemistry does not require a fresh mind as do physics and maths. If chemistry is attempted first, it leaves the two heavy subjects P and M one after the other for the already tired brain. Second, at the start there is no hurry, as we feel there is a lot of time left. So we would tend to spend more time than required in the inorganic and organic part in unnecessary thinking. I recommend starting your paper with P or M, whichever you are better at between the two.

For example, if your physics is better than maths, then start with physics. Starting with a better subject will increase your confidence and hence will be psychologically a better choice. Then take chemistry. Taking C in between gives your brain breathing space between the other two calculation-intensive subjects.

Then take Maths in the end in a relatively peaceful condition as two of the subjects have been attempted. If your M is better than P, then the recommended order of attempt would be M, C, and then P. In the second round, you can attempt the questions in any order, depending on how many Qs are left in each subject.

Combining 50-30-50 with the order, the paper would be attempted as follows (time is mentioned in brackets in minutes):

- If P better than M: P (50), C (30), M (50) Second Round (50)
- If M better than P: M (50), C (30), P (50) Second Round (50)

This strategy can be applied to both the JEE Mains and the Advance Exam. This strategy has shown marvellous results for many students after they practised it in 4–5 tests and got comfortable with it. If it does not work for you, then feel free to drop it and devise your own strategy.

Identify the Traps

Strategy 5 Know that there might be some Qs laid as a trap for us! They will challenge us—"Come on, fall into me," but never listen to them ("Dil pe mat lena"). *A trap Q* is a question that we might be able to solve but will take a lot of our time as compared to their worth of marks.

For instance, in JEE 2016, one paper had a total of 186 marks to be completed in 180 minutes, making 1 minute worth 1 mark approximately.

Now, suppose there is a tedious Q, which might take 15–20 minutes to solve, but will fetch only 4 marks. Is it worth doing it in the first round? The answer would be a 'No' if there are unseen questions still remaining. So once we identify a trap question, it is recommended to leave it and mark it for the second round.

We might be able to solve all the questions of the JEE, but solving them within the given time limit is what makes the JEE challenging. That is where the selection of the right questions to solve has its role. This selection becomes much more important when the paper is tough and the cutoffs are going to be very low. An example of such a trap Q is shown in Fig. 5.1; this problem appeared in physics in JEE-Advance 2016.

Two thin circular discs of mass m and 4m, having radii of a and $2a$, respectively, are rigidly fixed by a massless, right rod of length $l = \sqrt{24a}$ throught their center. This assembly is laid on a firm and flat surface and set rolling without slipping on the surface so that the angular speed about the axis of the rod is ω. The angular momentum of the entire assembly about the point 'O' is $\overline{L}$ (see the figure). Which of the following statement(s) is(are) true?

(A) The magnitude of angular momentum of the assembly about its center of mass is $17\ ma^2\omega/2$

(B) The magnitude of the z-component of $\overline{L}$ is $55\ ma^2\omega$

(C) The magnitude of angular momentum of center of mass of the assembly about the point O is $81\ ma^2\omega$

(D) The center of mass of the assembly rotates about the z-axis with an angular speed of $\omega/5$

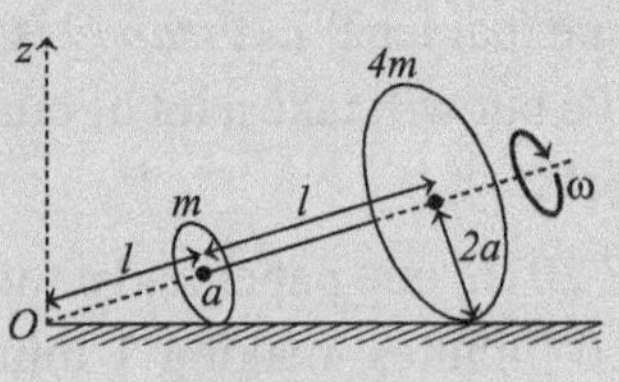

Figure 5.1 A trap question

In JEE Mains, leaving the tough questions for the end becomes much more important: it has more questions than JEE Advanced exam, which does not leave much scope for getting stuck in timeconsuming questions. The JEE challenges you intellectually, but do not approach it as a test of your intelligence only. Approach it as a smart test warrior. Be strategic. Be wise.

Guessing Answers

Strategy 6 In sections with no negative marking, we mark answers of all questions whether we know it or not because luck might work and there is no penalty for wrong answers! But the pain comes with negative marking. It can be tricky to take a risk in questions that we are not sure of.

In single-correct-answer-type questions, if we randomly tick an option, its probability to be correct is only 25%. You might not want to take such a big risk. Guessing an answer where you can eliminate at least two options is recommended. It takes the probability of a correct guess from the remaining two options up to 50%.

Now mostly in the JEE Main single correct questions, +4 marks are awarded for a correct answer and –1 for a wrong answer. If you guessed 4 such questions and if even one of them is correct, you are going to gain (+4 marks for one correct answer – 3 for wrong answers = +1).

If your fortune is good that day, more guesses may be correct!! However, avoid guessing answers in cases where you are unable to reject any option prior to the guess.

OMR Sheet Filling Strategy

In 2017, the IIT-JEE committee announced the JEE Advance 2018 exam to be conducted as a computer-based test (CBT) only. Strategies for attempting the JEE Advance CBT are discussed in the next chapter. Here, we will discuss strategies for those exams that are conducted in the offline mode, in which an OMR sheet is to be filled.

Once, after a practice test, one of my students came to me and said, "Sir, I solved nearly 80% problems correctly but got only 30% marks in the test." Surprised, I asked "How?" He explained: "I had decided to fill the OMR sheet in the end. I got deeply involved in solving the problems and lost track of time. In the end, I had only 5 minutes left and my whole OMR sheet was empty. I panicked and started filling it. When I had barely filled half of my answers, the bell rang and the invigilator took away my OMR sheet! To add to this, in the hurry, I filled some of the answers incorrectly." He learned his lesson the hard way. Many candidates, who adopt to fill the OMR sheet in the end, face a similar situation.

Strategy 7 NEVER leave filling the OMR sheet for the end, NEVER. You would never want to take such risk in the final exam.

An alternate way can be to fill each answer as soon as a question is attempted. In this way, the risk will be mitigated. Please ensure that you mark your answers in the question paper also along with the OMR sheet. This will be helpful for future reference. Another way can be to first attempt 10 questions (10 to 15, as per your comfort level), mark their answers in the question paper and then fill the corresponding bubbles of these 10 Qs in the OMR sheet in one stretch. The advantage of this strategy would be that after solving 10 Qs, your mind will get a chance to relax while you fill the bubbles.

Follow this for the initial 130 minutes of the first round (as part of the 50-30-50 strategy) and, in second round, fill the answer of each Q as you attempt it. Do not wait for 10 Qs, as the time allotted may end before you complete the process.

You should try both of these strategies and choose the one that best suits you. But NEVER leave the OMR sheet unfilled till the end!! Bubble it now!

The strategies mentioned in this chapter will benefit you after some practice. Try them out and retain those that work for you. But do practice them for best results. Practice is the key.

Scan the QR code to listen to the author on this topic

06

Going Online

JEE, known for surprises, in 2017 announced that the JEE exams will be conducted only as computer-based tests (CBT). What that meant was, instead of the questions being given printed on physical papers, they would appear on a computer screen. And answers, instead of being filled as bubbles on the OMR sheet, had to be marked on the computer itself.

One angle of looking at this could be: This change is to be faced by every candidate. So, why not take this change as an opportunity to get some lead. Let us understand the dynamics of CBT and become proficient in it!!

CBT, being computer based, is different from the OMR-based exam. The things which are favourable for a candidate in a CBT over an OMR-based test are as follows:

1 In the OMR-based exam, filling the bubbles used to take time, which will be saved in the CBT mode, as answers will now be marked with a click of mouse along with each Q.

2 In the JEE, the bubbles had to be filled using a pen on the OMR sheet. Answers once marked could not be changed. In a CBT, before final submission, the answers can be changed as many times as required.

3 While filling the bubbles, there was always a chance of filling the bubbles of the questions in the wrong order. For example, filling the answer of Q18 in the bubble of Q19, as the OMR sheet has bubbles one after the other. Many candidates have made this mistake, which gets sequentially cascaded in the following Qs till they realize. However, now in CBT, this hazard has been arrested, as the answer to each question is to be filled along with the question.

4 At any point of time, we can see the overall attempt status of the paper in a CBT—how many Qs have we answered, how many have been left unanswered, and how many have been marked for review, etc. If we needed this overview in an OMR-based test, we would have to manually count each question from the OMR or question paper.

There is one more advantage that you might get as a side benefit CBT means computers are required. To have this facility, the centres of an online exam are usually computer labs of universities (or similar places) that generally have air conditioners. And the JEE is held in summer time!! So, you may have ACs to beat the heat.

As candidates are familiar with the pen and paper format of examinations from school time, be it school exams or Board exams, initially some of the candidates are not comfortable with the CBT mode. Things that candidates find unfavourable in CBT over physical question papers are as follows:

1 Diagrams in the physical question paper can be directly used. Data, arrows, etc. can be scribbled on them and some of the questions get solved in the process. In CBT, the diagrams to be used in solving questions are to be re-made on rough sheets provided, which consumes the candidates' time.

2 Candidates have to juggle between pen and mouse. To scroll on screen, the mouse is to be used and to solve or write down some data, a pen is used. Sometimes data gets copied incorrectly in this juggle.

3 Pressure of the ticking timer: In the CBT mode, the timer is continuously ticking on the screen, which increases pressure for some of the candidates.

4 Looking at the computer screen for three hours is strenuous for the eyes as compared to the physical paper.

Now these conditions need to be managed optimally. Regarding the remaking of diagrams and rough work of the questions, we must strategically and wisely use the rough sheets provided. The time consumed in making the diagrams and solving the questions has to be invested wisely, so that while revisiting the questions we can reuse the diagrams we made and the calculations we did during our first attempt. Generally, the case with rough work is, as things are done randomly here and there, the calculations of the questions cannot be found while revisiting them, and they have to be done once again.

Rough work too can be done in an organized way. Rough sheets can be numbered and divided into parts as shown in Fig. 6.1. Writing the Q number along with the calculation/solution of a question would save time in case we have to revisit that question. It is not that we have to keep our rough sheets very neat, clean, and decorative. It is simply organizing them a bit so that we can refer to our work whenever required.

Q1 …	Q3 …
…	…
Q2 …	Q4 …
…	…

Figure 6.1 Organizing rough sheets

Also, many candidates have this query: What if additional rough sheets are needed? Feedback from candidates who have appeared for the JEE Mains online exam in previous years is that additional rough sheets are provided upon asking from the invigilator. If this is the case with the JEE Advanced exam also, do not hesitate to ask for additional rough sheets when required.

Proficiency in CBTs can be achieved through practice. The more you take practice tests in the CBT mode, the better it is. It will make you comfortable with CBTs, while familiarizing you to the computer screen, simultaneous use of mouse and pen, continuous sitting of three hours, and also to the ticking timer. If you are preparing by yourself or your coaching centre does not have the CBT facility, I would recommend joining some online test series which will give you some practice tests. Keep note of the time through the timer of the test and not a personal watch/mobile. This will train your mind as per the countdown timer and make you comfortable with it.

NTA generally releases some mock tests for JEE. These tests are very helpful, and I would say these are must-do to familiarize you with the online mode of the exam. Spend some time taking mock tests and exploring every possible thing that comes to your mind regarding the CBT. Also, candidates have a bit of insecurity about technical glitches in the CBT, as machines are involved. However, as reported by some of the students who have taken the JEE exam, in case of any malfunctioning of the computer, the candidate will not lose any of his/her test time. When the candidate gets back to the test portal, the timer will start from where the candidate left it at the previous instance.

Regarding the test strategies in a computer-based test, the strategies discussed in the previous chapter (other than the OMRfilling strategy) are applicable here also, as only the medium of the exam has changed. According to the mock tests issued for JEE, there is a facility to mark questions for review. You can see the questions marked for review in a different colour. This facility can be utilized while solving the paper in two or more rounds; the questions that we would like to visit in the second round can be marked for review.

These are those questions which we leave in the first round (to be tried in the second round) or the questions we have solved but are not sure of and would like to revisit in the second round. In the second round, we will directly jump to the questions that we had marked for review. As we solve questions in the second round, we mark their answers and clear their 'review' tag. After the second round, if time permits, you may go for a third round!

Before the end of test, make sure that you have marked the answers of all the non-negative-marking questions, i.e., those in which the wrong answer has no penalty. Mark them all. There may be some lucky marks in store for you!!

Let me share one specific experience of some of the candidates regarding 'mark for review' questions during their practice tests. While solving questions in the first round, candidates marked all the questions for review. Only those questions were not marked for which they were perfectly sure of the answer. This increased the count of review questions.

The marked Qs included those Qs also which the candidates knew were out of their scope (which means they knew that they could not solve them; yes, there are such questions!). So, when they attempted the questions marked for review in the second round, they wasted time in going through the out-of-scope questions they had marked as well, leaving less time for questions which were in scope and could actually be solved.

In such cases, the strategy that can help these candidates is: They should not mark the out-of-scope questions for review. They should leave such questions in the unanswered category. Now, after the second round, when all the review questions have been attempted, they can go through these unanswered questions if time permits. Based on the mock test given on the JEE website for JEE Advanced-2018, Fig. 6.2 presents a snapshot of the mock test with some pointers marked on it.

Taking inputs from the mock test, the utility of each pointer is listed. It would be great if you could open a mock test on the JEE website as per your JEE year and explore it along with going through the following section. This will enhance your understanding of the CBT.

These inputs are as per the year 2018, I would recommend you to see the Mock tests as per your year of JEE attempt.

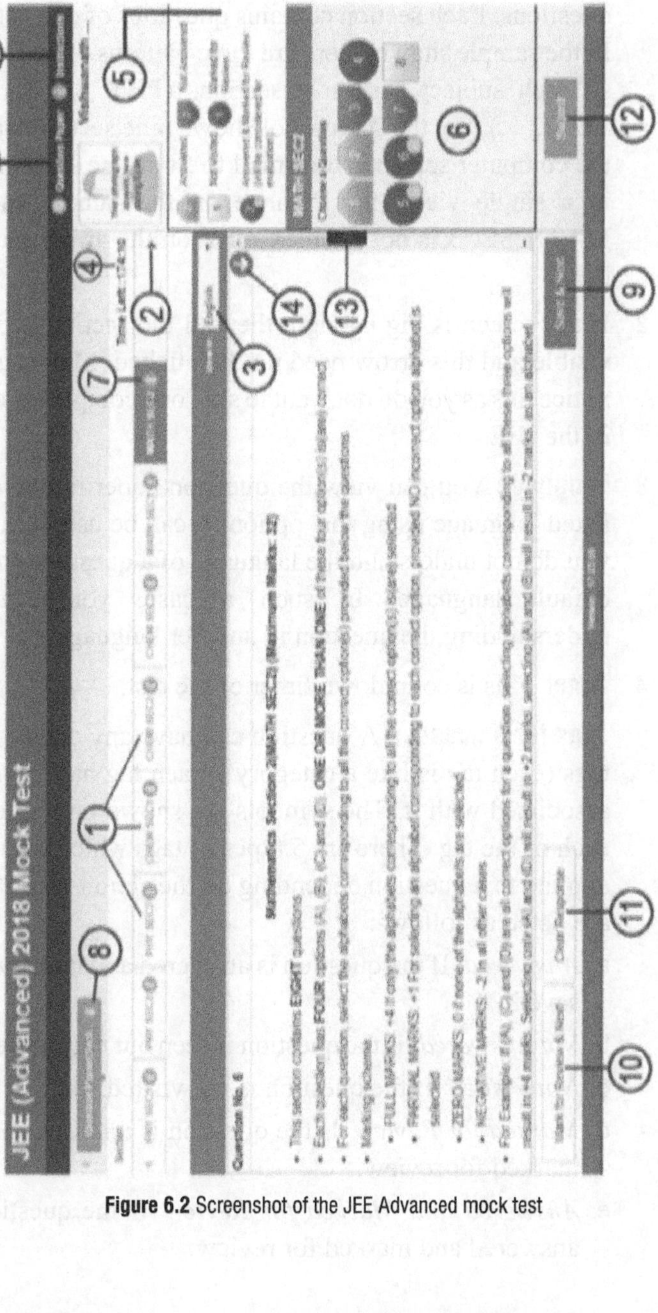

Figure 6.2 Screenshot of the JEE Advanced mock test

1 **Sections** JEE Advanced has different patterns of questions. Each section contains questions of one pattern. In the sample shown, there are three patterns of questions, so each subject has three sections (PHY SEC1, PHY SEC2, ...). So, the screenshot shows nine section tabs. If the computer screen is too small to fit all the section tabs in a single view (for instance, in the sample shown, MATH SEC3 is not visible), click on the arrow marked by ...

2 If the screen is big enough, then all the sections will be visible and this arrow need not be clicked. Make sure to notice this as you do not want to skip one complete section in the JEE.

3 **Language** You can view the question paper in any other listed language using this option. It can be useful in case you do not understand the language of a question in your default language. In such a case, you can try understanding the question in another language.

4 **Timer** This is countdown timer of the test.

5 **Tags for a question** A question can have any one of the 5 tags (each tag is like a category). Each tag has a symbol associated with it. The symbols are shown on the left of each of the tag. There are 5 types of tags which would be applied to a question depending on the action done. They are listed as follows:

a. *Answered* If the question is answered and its answer is saved.

b. *Not Answered* If the question is seen but not answered.

c. *Not Visited* If the question is not visited yet.

d. *Marked for Review* If the question is *unanswered* and marked for review.

e. *Answered and Marked for Review* If the question is answered and marked for review.

6 **Section panel** This portion has the numbers of questions in the section one is currently attempting. For instance, in the sample shown, we are in MATH SEC2, which has eight questions. Now each question has a symbol of the tag on it. In the sample, Qs 1 and 2 are 'Answered', Qs 3 and 7 are 'Not Answered', Q8 is 'Not Visited', Q4 is 'Marked for Review', and Qs 5 and 6 are 'Answered & Marked for Review'. Notice the additional box in the symbol of 'Answered & Marked for Review' to differentiate it from 'Marked for Review'.

The interesting thing is, in area 5 the number of questions shown in each tag gets automatically counted and displayed. For example, Qs 1 and 2 are answered, so two questions out of eight in this section are answered, so notice a 2 written on the symbol of the 'Answered' tag in area 5. Similarly, two questions (Qs 3 and 7) are 'Not Answered', and so on.

7 **Section Summary** Taking the mouse to the ⓘ symbol in the section tab gives the tag counts of the questions from that section. So, whenever you want to see the status of some section, you need not go to that section. Just take the mouse to the ⓘ symbol of that section and you will see the summary.

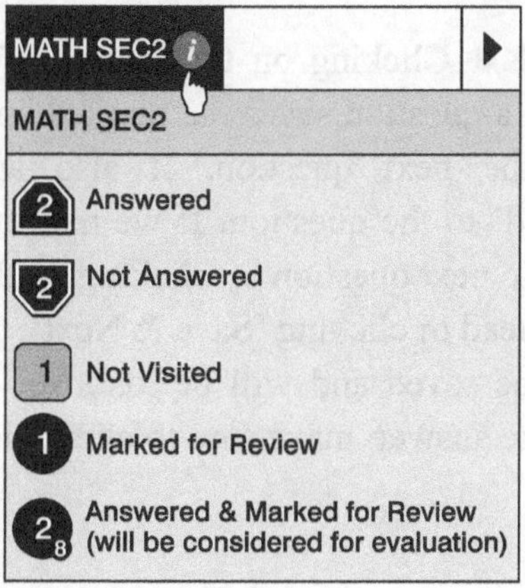

8 Overall Summary Taking the mouse to the ⓘ symbol in the Question Paper tab gives the tag counts of the questions from the whole paper. This will give you the overall statistics of the status of questions of the full paper. In the sample shown, 17 Qs were 'Answered', 11 Qs 'Not Answered', etc. out of 54 Qs in the question paper till that point of time.

9 Save & Next Clicking on this button after marking an answer in a question saves the marked answer and takes you to the next question. It also applies the tag 'Answered' to the question. If we mark the answer and move on to next question by clicking on the Q number in area 6 instead of clicking 'Save & Next', then our answer will not be saved and will be lost. So, once you have marked the answer, make sure to click this button.

Even if you want to leave the question unanswered, then also the 'Save & Next' button will take you to the next question, keeping this question unanswered and applying the tag 'Not Answered'. So, click this button after attempting every question unless you want to mark it for review.

Now one more thing to notice is, there is no separate button for going to the previous question like there is for the next question. If you want to go to the previous question or some other question, click that particular Q number in the area 6.

10 **Mark for Review & Next** If you have marked answer of a question then clicking on this button will save the answer, apply the tag 'Answered and Marked for Review', and take you to the next question. If you have not marked the answer of a question, then clicking on this button will apply the tag 'Marked for Review' to this Q and will take you to the next Q. This button basically applies one of the two review tags depending on whether or not the answer is marked.

Suppose you are revisiting a question that is 'Answered & Marked for Review' or just 'Marked for Review'. Now you are done with that Q and want to remove the review tag. Click on the 'Save & Next' button. This will remove the review tag and apply the 'Answered' or 'Not Answered' tag depending on whether you have answered the Q or not. This will also save your response.

11 **Clear Response** Clicking on this button will clear the marked answer(s). It will also remove the original tag of the question and apply the 'Not Answered' tag.

Clicking on this button will keep you on the same Q. Thereafter, the tag will be changed depending on what action you take.

12 **Submit** Now this is the button you need to be cautious with!! Clicking on this button will prompt you to submit your answers and end the exam. If by mistake you clicked on this button, do not panic. A window will appear which will reconfirm if you want to submit the answers for marking (it is demonstrated in the mock test). Clicking on the 'No' button will take you back to the test. If you have finished your exam and want to submit it, then click this button and proceed as per instructions.

13 **Arrow** Clicking on this arrow hides the question number and tags panel. This is to view a question in a wider area. Clicking back on the arrow will bring back the panel.

14 **Arrow** Clicking on this arrow scrolls the view to the end of a question and clicking on the upward arrow scrolls back to the start of the question.

15 **Question Paper** If you want to see the full question paper in a continuous mode, click on this button.

16 **Instructions** For viewing instructions during the exam, click on this button.

These inputs are based on official mock tests uploaded for JEE Advanced-2018. The explanations are based on my exploration of the mock tests and help video. It is strongly recommended to rigorously explore the mock tests of your JEE year for better understanding of the functionalities. You may discover something more and this process will familiarize you to the test platform.

Also, please note that these mock tests are for the year 2018. Things may (or may not) change in the years to come. So, I sincerely request you to first have a look at the JEE website before drawing inferences from the snapshots and inputs given in this chapter.

Scan the QR code to listen to the author on this topic

07

You Can Do Better

Post-test Analysis

What is the purpose of the tests that you take as practice before the actual JEE exam? To embarrass yourself?! Not at all. These are meant for you to know your strengths and weaknesses. They act as consultants for you, provided you give them a chance. Most of the students start calculating their marks as soon as the test is over and eagerly wait for the results. And after they know their ranks (how much they have scored), they forget about the test, and move on.

That is natural. However, what gets missed is the detailed analysis of the test. Without detailed analysis, a test cannot serve you any more than regular homework done at home.

One of my students (who got into IIT Kanpur) called me just after his JEE exam. In an excited tone he told me, "Sir, I cracked two of the tough Qs very easily thanks to the test analysis technique. These two Qs were very similar to those I had come across in one of my practice tests, and I had done them incorrectly then. After the test, I analysed it, wrote the correct solutions, and noted my mistakes in those questions in the test analysis copy. Sir, when I saw those two Qs in the JEE, the image of my test analysis copy flashed in my mind and I was jumping with excitement, as I knew the correct approach now and also the mistakes to be cautious of. These two Qs gave wings to my confidence."

Two correct Qs in JEE-2016 meant 8 marks, which was the difference of marks scored by AIR-3000 and AIR-4000 (AIR around 6000 and 7500 scores also differed by just 8 marks). On the contrary, if he had not bothered to do the test analysis, the same two Qs would have adversely affected his confidence due to a feeling of guilt in the middle of the exam. Remember, there is no failure, only feedback, if we would learn from it.

How to Analyse your Tests

Make a separate notebook for test analysis. After each test, *within a week*, analyse each question as follows. For each Q, any one of the three possibilities (correct, incorrect, not-attempted) can occur:

1. Correct answer: If you attempted a Q correctly, then write *Correct*, and if a better solution than yours exists, note it down.
2. Incorrect answer: Write *Incorrect*. Note down its correct solution. Write down the exact mistake that you made (with a different colour pen or highlight it). Do not just write 'silly mistake'. Be as precise as you can, because when you will revisit the analysis, mistakes should be clearly recognizable.

3. Not-attempted questions: Qs of which you have not marked the answer in the OMR sheet. Write *Not Attempted*. Note down its correct solution. Most important is writing down the exact reason (highlight it) why you left it: Lack of time, do not know the solution, long question, not sure of the answer, fear of the topic, etc.

Test analysis will take some time, but it is an investment with great returns!

How to Use the Test Analysis

"Mistakes are for *learning*, not *repeating*." Before every test, review all the mistakes (highlighted) of the previous tests that you have written in the test analysis notebook. This will ensure that you do not repeat the same mistakes in upcoming tests, and most importantly in the JEE. I still remember one of the mistakes I made in my first practice test in Class 11. It was a question on vectors where 'magnitude of change' of $\hat{i}$ and $\hat{j}$ was asked and I understood it as 'change of magnitude'. So, instead of $\sqrt{2}$ as the answer, I marked 0, and got 0 marks for it!! From then onwards, whenever the words 'magnitude' and 'change' came together, I used to become alert and read them extra carefully.

Second, from the compiled analysis, identify the pattern of mistakes you are making. It will lead you to pinpoint the problem areas that can be cured effectively. For example, you might notice that calculation mistakes are hurting you badly or you are not able to attempt all the questions due to your speed or the Qs of some particular topic are bothering you. For example, 'location of roots' in the Quadratic Equations chapter is problematic. Without post-test analysis, you would have labelled Quadratic Equations as a *weak chapter* and started revising it from the beginning, whereas what was needed was concentrated effort on the specific topic 'location of roots' in that chapter.

After analysing your test, you will get to know the problem areas that you need to focus on and thus your productivity will increase. JEE is all about being a smart test warrior. A warrior is intelligent and strategic. *Embrace the right strategy and win!*

Silly Mistakes

I used to make many silly calculation mistakes in the beginning of my 11th standard. It hurts really bad when we know the correct solution of a question but we lose marks due to some silly calculation mistake. To get rid of these mistakes, I started devoting 15 minutes daily just for calculations. I used to take random calculations (such as 537×29, $730 - 356 - 47 + 48 \times 2$, etc.) and solve them. Along with this, I minimized the use of the calculator in question solving and started approximating calculations. Initially, it was difficult to get rid of the calculator, but soon when the correct answers started coming, the process became enjoyable. To my surprise the process worked well and reduced my calculative mistakes significantly!!

In the JEE, the sad thing is that a calculator is not permitted, but the good thing is that most of the questions are smartly designed to not have tedious calculations. So, it is wise to try (and try hard) not to use a calculator while solving questions, particularly previous years' JEE questions. Look out for approximations, cancellation of numbers, and rounding off digits, and you will be pleasantly surprised upon getting to the right answer on your own.

Remembering

1 Squares till 25

2 Cubes till 12

3 Fraction till 1/10

is very useful, particularly for *integer type questions* (where the answer is an integer from 0 and 9 , both inclusive). These things, if memorized both forward and reverse, remarkably improve the speed and ease of calculations. You can write these on a paper and put them up in your room.

1^2	=	1	1^3	=	1
2^2	=	4	2^3	=	8
3^2	=	9	3^3	=	27
4^2	=	16	4^3	=	64
5^2	=	25	5^3	=	125
6^2	=	36	6^3	=	216
7^2	=	49	7^3	=	343
8^2	=	64	8^3	=	512
9^2	=	81	9^3	=	729
10^2	=	100	10^3	=	1000
11^2	=	121	11^3	=	1331
12^2	=	144	12^3	=	1728
13^2	=	169			
14^2	=	196			
15^2	=	225			
16^2	=	256	1/1	=	1
17^2	=	289	1/2	=	0.5
18^2	=	324	1/3	=	0.33
19^2	=	361	1/4	=	0.25
20^2	=	400	1/5	=	0.2
21^2	=	441	1/6	=	0.167
22^2	=	484	1/7	=	0.143
23^2	=	529	1/8	=	0.125
24^2	=	576	1/9	=	0.11
25^2	=	625	1/10	=	0.1

Examination Phobia

The word 'exam' itself can give many people a feeling of butterflies in their stomachs, and during the examination, the butterflies convert into fire-spitting dragons! Anxiety (up to a certain level) before an exam is natural and sometimes acts as a favourable factor. You become anxious to perform well and it prompts you to study more. However, beyond a limit, anxiety becomes an obstacle in your performance.

During the examination, it leads to committing silly mistakes, forgetting formulae, and sometimes blackouts. The same Q that was a piece of cake at home looks frightening in the exam. *Anxiety is the fear of a future that has not happened.* There are ways by which you can remove this illusory fear from your head.

How to cope with and overcome anxiety Take a test once a week at home, simulating the exact exam environment. If you are taking a 3-hour test, set your timer for 3 hours or use a clock, keep your mobile away (this is the hardest part—do not use the clock of the mobile phone, as this will not be the case in the actual exam), get cut off from the outside world for those 3 hours, and focus only on the test paper kept on your study table. Believe me, this process will familiarize you to the butterflies and they will become your friends. Try it out and discover for yourself!

Speed versus Accuracy

"My speed in the tests is too less, some Qs are getting left unattempted," "My accuracy is less, I receive many negative markings." These are very common problems faced by candidates. If they go for speed, accuracy goes down and *vice versa*. Well, speed and accuracy are tricky to handle together.

If your speed is slow in the tests, you might not be picking the right Qs. The test strategy of 50-30-50 (described in Chapter 5) can be useful for you. Also, in some of the practice tests that you will take at home (as discussed in the previous section), keep your focus mainly on the speed (and do not worry much about accuracy). If you cannot spare three hours for a test at home, then take tests of smaller durations (say, one hour).

Select the number of Qs appropriate for one hour out of your homework or revision and attempt them within the time limit of one hour. For instance, if you want to increase speed for the JEE Mains pattern, then take 30 Qs of level similar to those in the JEE Mains pattern and set the timer for an hour (as the JEE Mains has 90 Qs for 3 hours). Practice this twice a week.

Your aim should be to attempt as many Qs as you can within the time limit. After the time gets over, do not forget to attempt the Qs that you left during the test. Also analyse which Qs took you more time to solve and where your decisions of attempting or leaving a Q were right.

Following this process for some weeks will increase your speed.

Caution Avoid this 'speedy attempt' method with the homework of the current topics that you are studying for the first time. Such topics should be given their due time for grasping the complete depth of the concepts involved (as you have just started understanding them). Do not set time limits for them; spend ample time to absorb the concepts first.

'One or More than One Correct' Type Questions

Often, one or more than one of the options becomes wrong while attempting 'More than One Correct' type of objective problems. If you are facing this problem or your confidence shakes while attempting this type of problem, try the following method, which has helped a lot of students.

Method While practising at home, do five 'More than One Correct' type of Qs in a stretch *without* seeing their answers/solutions, and after you have attempted these five Qs, then crossverify your answers with those given in your book. Normally students are tempted to see the answer as soon as they attempt a Q.

Unknowingly, this temptation turns into a habit of looking at answers as soon as they finish reading the Q and then finding the logic to justify the answers given. So the mind is being trained to find explanations after seeing the answer.

But alas! The test paper does not have answers at the back of it. So let us practise doing five Qs without seeing their answers and train the mind as it is required. The temptation of looking at answers beforehand is very difficult to overcome, but I am sure you can do it.

Test analysis is necessary to gain most out of your tests. It helps in identifying the weak areas precisely and avoiding silly mistakes. Practising tests at home will improve your speed as well as accuracy and help you cope up with examination phobia, if you have any. You can take subject-wise tests for improvement in the respective subjects (PCM). Each subject has its own nitty-gritty. The approach to study each of them, along with maintaining a balance among the three, is discussed in the following chapter.

Scan the QR code to listen to the author on this topic

08

Subject-Wise Tips

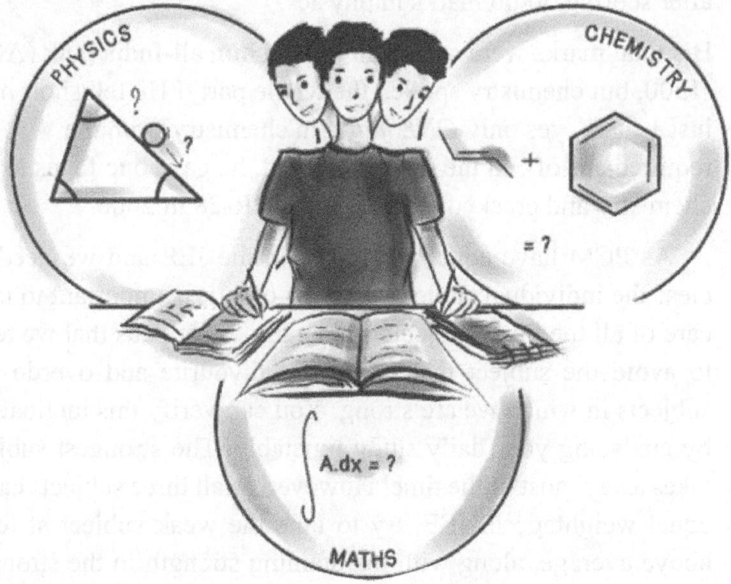

"I love maths and physics but chemistry smells like rotten egg H_2S gas to me," "Maths is too scary to handle," I score well in chemistry, but physics marks are a nightmare," etc. This is common chit-chat among JEE aspirants.

Managing the three subjects physics, chemistry, and maths (PCM) is an art needed for the JEE and it can be learned. You might be in love with one of the three subjects, which is perfectly normal. This loved one brings marks for you and your love keeps on increasing. However, due to this loved one, you should not have a *break-up* with the less liked ones (or hated ones). Individual subject-wise cut-offs are to be achieved for all three subjects in the JEE.

One of my IIT Bombay (computer science) batchmates, Shashank Singh (now one of the finest software engineers in the US), could not clear JEE in the first attempt in the year 2005 even after scoring 50/60 marks in physics.

His total marks were sufficient to land him all-India rank (AIR) -1500, but chemistry spoiled the whole party! He fell short of just 1 mark, yes only *ONE mark*, in chemistry from the required cut-off. In the second attempt, he gave due focus to chemistry and cracked the JEE with AIR-28 in 2006.

As PCM have equal weightage in the JEE and we need to clear the individual subject-wise cut-offs, it is important to take care of all the three subjects. Generally, it happens that we tend to avoid the subject that is not our favourite and overdo the subjects in which we are strong. You can verify this inclination by analysing your daily study timetable. The strongest subject takes away most of the time! However, as all three subjects carry equal weightage in JEE, try to take the weak subject at least above average, along with maintaining strength in the stronger ones. For a subject that you do not like studying or that is your nightmare (say, for example, maths) and your dearest wish is to remove that subject from this planet, I would suggest trying the following. First of all, write in big letters on a piece of paper 'I LIKE Maths, in fact I LOVE it' and put it up in your study room where it is frequently visible. Reading it many times will make your mind start believing it. Bingo! Now when you pick up maths, you are no more afraid of it. Try this out. It has worked for many candidates and now it is your turn.

Also, when students declare a subject as a weak subject, or some chapter as a weak chapter, in most of the cases they are doing injustice to that subject/chapter. If we carefully examine the situation, the whole of the chapter is rarely the problem, leave aside the whole subject. It is just some aspects of that chapter that we are having difficulties with and we have not identified them yet. It is like, if someone gets a bone fracture, does it mean they have a fullbody fracture? Will any sensible doctor plaster the whole body?! No!!! First of all, an X-ray will be done to identify the exact location of fracture and then the fractured part will be plastered. Similarly, we also need an X-ray (test analysis) for identifying fractured parts and apply the *plaster of practice* at the exact location. After some time, we will be fit and fine, ready to rock!!

Each subject has its own characteristics, and a unique toolkit is required to handle each of them. Let us look at these one by one.

Physics

> ***All of physics is either impossible or trivial. It is impossible until you understand it, and then it becomes trivial.*** —*Ernest Rutherford*

Physics is the subject of concepts. The MANTRA to make physics yours is 'chase the concept and its questions will follow'. While solving a question in physics, look for the underlying concepts; do not just blindly run behind solving as many Qs as you can. Spend time on each problem to understand the fundamentals and concepts involved. After reading a question in physics, visualize the problem, identify the concepts to be applied, make a diagram (if needed), and then write the formulae/equations to solve the problem.

Along with the concepts, there are laws and formulae to be memorized. I would recommend creating a *self-written* 'formula sheet' for each chapter, which has all the relevant formulae of that chapter. Whenever you forget a formula, refer this formula sheet. Also, when a quick revision of some chapter is needed, just go through the formula sheet. If you are having problems in remembering the formulae of some topic, paste its formula sheet in your room for some days, where it is frequently visible, and see the results. Try it out!!! This can be done for all the three subjects.

Performing experiments in the physics lab is helpful in better understanding and visualizing the concepts. For example, getting hands-on experience with instruments such as the vernier caliper, screw gauge, potentiometer, etc. will give you an upper edge in solving unconventional problems asked in the JEE on these instruments.

Try relating concepts (particularly in mechanics) with your everyday experiences such as cycling (how tyres rotate, paddling, gear mechanism—if you have a geared bicycle), floating something in water, friction, and force. Feel them and you will fall in love with physics.

Caution Physics might tempt you to go into concepts that are far beyond the requirement for the JEE and you may end up investing extra hours at the cost of the other two subjects. I would recommend sticking to the syllabus for the sake of the JEE, as the time available is limited. Being goal-oriented is the key.

Chemistry

Chemistry deals with reactions, colours, smell, names, chemical formulae, equations, etc. In chemistry, memorizing power plays a significant role along with understanding concepts and solving numericals. Chemistry has three verticals: physical, organic, and inorganic chemistry.

Physical Chemistry

This part of chemistry has a blend of chemical systems and the principles of physics. It has many interesting concepts to understand, numerical to solve, and formulae to be remembered. You could make a formula sheet having all the formulae and important points of a chapter and refer to it during revision.

The numericals that you will solve will have calculations, which will sometimes be with tedious data. It may tempt you to employ a calculator for quickly getting to the answer, but be cautious. Avoid using the calculator as much as you can as it is not allowed in the JEE. Practise manual calculation, especially while solving previous years' JEE Mains and Advance questions. Most of these questions are designed cleverly, such that the calculations can be manually handled. With practice, you will become good at calculations and you will be amazed to see yourself cracking tedious calculations on your fingertips.

Organic Chemistry and Inorganic Chemistry

Most of the Qs in organic chemistry (OC) and inorganic chemistry (IOC) are non-calculative and theory/information-based, which can earn some quick marks for us and save a lot of time in the exam. I would like to call these subjects *time savers and quick gainers.* There are theories and concepts, but what is different in OC/IOC is that a lot of things like names, colours, and reactions have to be remembered. And yes, not to forget the exceptions!

When there is a lot of information to be remembered, *mind maps* prove to be a very useful tool. A mind map is a way of organizing information in a non-linear way in the form of a chart. For example, the reactions of Fe+3 ion can be mapped as shown in Fig. 8.1. You can colour the compounds in their characteristic colour. This will create a long-lasting visual memory.

However, do not forget to write the names of the colours besides the compound as illustrated. Mind maps can be made for various reagents and groups such as alcohol, ketone, etc. These charts have the consolidated information/reactions of a group at one place. Mind maps will help you to remember a lot and for longer.

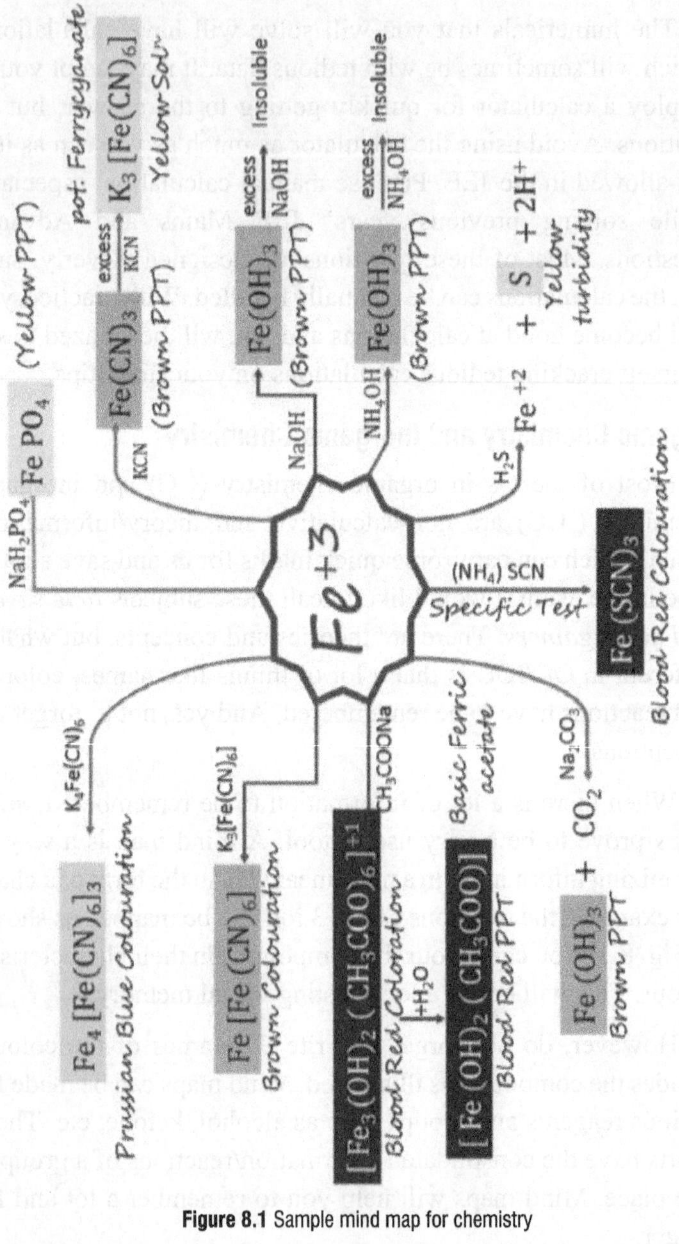

Figure 8.1 Sample mind map for chemistry

To make the learning process interesting, peer learning can be employed. After you and your friend have revised a chapter, play a quizzing round of asking Qs from that chapter turn by turn. This process will create stronger memories, as it involves many of your senses. Also, you and your friends will not let each other fall asleep while memorizing!

For remembering the sequences of names, mnemonics can be handy. For example, Group 14 elements of the Periodic table (C Si Ge Sn Pb) can be committed to memory as follows:

Chemistry Sir Generates Sneaky Problems

There exist many variants of such mnemonics. You could invent your own variants. If you want to add some entertainment, try inventing mnemonics with your friends. Also, doing practical experiments in the laboratory and actually seeing the reactions happening with all the colours and smells forms vivid memories in the mind and helps in better remembrance.

As OC/IOC have a lot of information to be remembered, sometimes, before the exam, the feeling that we do not remember anything creeps into the mind!! These are phantoms of the mind. These phantoms tell you:

"You will forget everything; you will fail." Understand that they are just phantoms. The practice that you have done over the year(s) is stored safely in your mind. At the time of the exam, they will emerge unaffected. *Trust your work, not the phantoms*!

Mathematics

> ***The only way to LEARN Mathematics is to DO Mathematics.***
> *—Paul Halmos*

The key to mathematics is 'Do it'. The more you practise doing it, the better you get at maths. If you wish to become a master in maths only by reading solved problems, then you are going to be disappointed. Reading the solution of problems will give you a satisfying feeling or will get you started, but it is only by doing and solving problems on your own that will gain you confidence in mathematics.

For strengthening weak chapters in maths, you may start with revising their theory and then reading some solved examples. When you have understood the solved examples, redo the same examples again by yourself without seeing their solution. After achieving a comfort level, try attempting new Qs without seeing their solutions.

If you get stuck, do not see the whole solution. Just see the first line of the solution; use it as a hint and try cracking the rest of the solution. If you are not able to do so, see the second line of the solution and so on. This will help you streamline your thinking process and identify the missing concepts. Do not give up. Keep practising.

Identify the Profiles

In mathematics, there can be hundreds of questions in a chapter, but if you carefully observe them, the profiles or type to which each of them belong will be limited. Each profile has a method to solve it, and some profiles that look tedious actually have a particular trick to solve them. Focus on these methods/tricks.

So, when you start solving a problem, first identify the profile to which it belongs and apply the method you learnt to crack that profile, and the Q will be taken care of. Each of the three subjects are equally important for the JEE and need attention and practice. Appropriate subject-wise methodology will help you gain command over all three subjects. To achieve the goal in the specified duration, subject knowledge has to be supplemented with determination, wise management of time, and smart-hard work. The goal is waiting for you. Just follow the path to high performance.

Scan the QR code to listen to the author on this topic

09

The Path to High Performance

On Janmashtmi— the celebration of the birth of Lord Krishna— every year, thousands of people get together to celebrate *Dahi Handi.*

Dahi Handi is a re-enactment of the special trick that Krishna used to get to his beloved yogurt/curd (*dahi*), which used to be suspended at a height by his mother to prevent him from eating too much of it. He would form a pyramid of people and climb on them to reach the pot. This *Dahi Handi* is celebrated in a grand way throughout the nation, more so in Mumbai. Various teams, fondly called *Govindas* (one of Lord Krishna's 108 names), gather to break the pot and take the prize. Let me note down the various characters involved here.

1. The Pot
2. Govinda, who breaks the pot
3. People who form the pyramid
4. Prize for breaking the pot
5. Folks who throw water
6. Spectators

This story of *Dahi Handi* is the story of someone amazing, confident, hard-working, and smart! Before you start thinking about random people, let me tell you who that person is. I am talking about YOU! How? Let us compare the characters.

The Pot	Your Goal (an All India Rank)
Govinda, who breaks the pot	You
People who form the pyramid	Parents, teachers, friends, etc.
Prize for breaking the pot	Dream college
Folks who throw water	Distractions
Spectators	All other people such as relatives and acquaintances

Setting the Goal

Imagine there is no pot. What will you break? If you want to break a pot, you must first have a well-defined pot! Sounds simple, but many of us do not have a clear goal. When I ask my students about their goal, most of them say, "I want to get into an IIT or I want to crack the JEE."

But will you be satisfied with AIR-9000 in the JEE or selection into any random IIT to accomplish your goal? No? Then set your goal precisely. Second, is your pot visible? Is your goal a written-down one?? A goal that is written down and within sight helps you to focus and avoid distractions. The goal is the first step towards achieving your dreams. *What is your goal??*

Breaking the Cycle of Stress

In this journey, who is Govinda: The pot breaker? YOU! You are that one person, the only person, who will make you reach your goals. Remember that you are on top of the pyramid. This means that you have to climb the highest; your responsibility is the greatest. It also means that you have to take the greatest risk. You have got to be equipped with revision strategy, notes making strategy, test preparation strategy, stress handling strategy, and recreation strategy, among other key strategies. Remember, you can break the pot if you beef up your strategy!

First thing, make it very clear to yourself that this journey of one/two years will be tough. This is because of two reasons: One, it is a long stretch of time. So there will be ups and downs, challenges, sacrifices, and victories. Two, there is an insane amount of competition. This in turn makes victory amazing. Make sure that your mind knows and accepts that there will be stress. There are three type of stress:

- Examination-based
- Self-made
- Distraction-based

But don't worry, we have the weapon. The greatest weapon against stress is our ability to choose one thought over another. **Let us see how for the three cases.**

Examination-based Stress

We treat tests like war! Tests bring with them fear and futile anticipation. The first and the most important step to handle this stress is adequate preparation, which I have covered in detail in other chapters. The second step is to reduce the importance of the test a few days before the day of test.

Let me share an interesting story. Arpit Maheshwari, AIR-2 IIT-JEE 2006, realized that the tests were getting into his head early in Class 11. This was not letting him perform at his best. So he made a strategy. For full 24 hours before a test, he did not study the syllabus of the test! He studied other things.

For example, if you have NLM in the test syllabus, he took up some other topic such as projectiles. This relieved all the pressure! The second measure he took was to avoid friends who were superstressed and who would transfer all the tension to him. Just before the test, he surrounded himself with those friends who were relaxed.

Like Arpit, you too can find your own way to dodge test pressure. Do not make tests larger than life. Also, the company of friends you surround yourself with matters a lot. You might have experienced that you were calm before one of your classmate asked you *"Test ki saari tayyari ho gayi kya yaar?"* (Are you fully prepared for the test?), and then something triggers in you and calmness goes for a toss!

Remember, the tests are to prepare you for the examination! If you accept being afraid of the tests, then you are preparing for being afraid in the examination also. And when we are afraid, we underperform.

Self-Made Stress

Self-made stress is of two types, as discussed in the following.

Burnout stress

First is when you burn yourself out, when you do not take breaks. You have no recreation. You keep all your worries to yourself. You do not chill out strategically. This way you invite anxiety and stress.

Let us look at how to reduce this kind of stress. Understand that it is practically impossible to just only study for two years. Or even for a month for that matter. Two things are important here: recreation and interaction.

Recreation All of us know what makes us happy. Make sure that you do those things at regular intervals. Anything that makes you forget everything else will work. It can be playing some sport or going out for a walk or a movie or talking to friends. It will add fun to your journey and make it enjoyable. You will have to sit for long hours during these two years to study and prepare for the exams.

Make sure that you keep yourself physically fit. Remember, a healthy mind resides in a healthy body. Playing some sport or doing *Yoga* or running a mile once in a while will keep you active and energized throughout the day.

Interaction Humans are social beings. You are too. You will sometimes fail, have complaints, feel anxious, and what not. Do not keep these feelings buried inside you. Sometimes stress becomes overwhelming with all sorts of negative thoughts pouring into the mind. In such situations, just get up and talk to someone. Do not wait. Just get up from that place. Achieving your goal is a team process. Remember, to break the pot, you need people to form a pyramid. Identify your pyramid-forming people. Family, friends, and teachers can be an effective support system. They are there to help you. They understand you. They care for you.

Share your problems with them and watch their feedback. Do not hesitate in sharing your failures and fears with them. If you feel you are helped by their words and thoughts, continue to share more. If not, find another person till you have your perfect pyramid of people to climb upon and break that pot. What will you share today? Go on share it now...

Last Moment Stress

The second kind of self-made stress is when you wait till the last moment to start studying; and when you start studying without a plan. The bridge between you and your goals is planning. Refer to Chapter 10 to know more about time management and planning.

Distraction-based Stress

Distractions are like the *people who throw bucket full of water* at you. Distractions are of infinite types, or in IIT-Bombay lingo, *"Infi distractions hain yaar life mein!"* (There are infinite distractions in life.) This is the problem of all human life.

I will focus here on mobile/Internet, friends/people, and love/ relationships. I have noted that these three things are the major distractions in student life. Year after year, hundreds of students have come citing one/some of these as a reason for bad performance.

Mobile/Internet In today's age, all/most of us have smart phones. Social media, movies, etc. take a lot of time without us realizing it. Before you know, there go 2 hours in scrolling through your FB newsfeed or Whatsapp or in watching YouTube videos. If you think carefully, these things can kill time if not used properly.

Make it a rule that you will use your mobile for Internet usage only for a specific number of hours that you feel are right and strictly adhere to the code. I will suggest that you not use it for more than an hour everyday. The time limit we are setting is for non-educational things. This will save you a lot of time.

Another thumb rule is: Keep your phone on silent mode while studying. There is nothing more distracting than notifications. A lot of research has been going on on the distracting power of notifications. Tell your parents your study hours, so that when you do not pick your phone up while studying, they will not worry.

Friends/people There are people who have no goals, who waste time in unnecessary things, or people whose goals are not thesame as yours. For a period of these one/two years, filter your friend circle and be around those who serve you the best in the goal you have set for yourself. Your success and your failure is your and only your responsibility.

Sometimes friends come to your room and start gossiping or insist that you go out with them when you are studying. Saying "No" at that time is very difficult but necessary. If you say "No," the message is very clear. Your goal is your first priority. Going out with friends is not bad, but only if it is for a limited time and most importantly when you want to go, and not when you are forced to go. Your choices make you who you are.

Say "No" to procrastination today. Say "No" to distractions. Say "No" to unhelpful influences. Say "No" to bad habits! When you can do this, you can finally say "Yes" to your dreams.

> ***Take up one idea. Make that one idea your life—***
> ***think of it, dream of it, live that idea.***
> ***Let the brain, muscles, nerves, every part of your***
> ***body, be full of that idea, and just leave every other***
> ***idea alone. This is the way to success.***
>
> *—Swami Vivekananda*

Love/Relationship All of us are hormone-driven, especially, at your age. So, all the feelings that you go through are natural and undoubtedly very powerful. You will see people around you in relationships (you know what I am talking about). Nowadays movies have made being in a relationship trendy and we perceive ourselves as cool.

Relationships are complicated. Sometimes you may like one person and sometimes your mind wanders from person to person. Just like the one-to-many function in mathematics! Beware!!!

Relationships need commitment. So do IIT JEE exams. Do you have the skills to effectively stay put on your IIT goal while managing a relationship? Understand that you have to be very honest in your assessment. If you find that the relationship is not supporting your goal, you have to make the tough choice about which path you want to proceed on. The choice is yours!

Sometimes the 'spectators' (your relatives or acquaintances) have expectations from you, which can become a source of distraction and unnecessary pressure. But like *Govinda*, you need to just focus on the pot and your teamwork. If your focus is shifted to the spectators, you may fall. Remember, your reward is in the pot.

Peer Learning

To reach the pot, you need to work as a team with your friends. Discuss, make strategies, and learn tactics from each other. Yes, peer learning is highly beneficial for achieving great results. During my preparation days, I used to sit with my friends for hours and solve problems together. It used to boost my energy, motivate, and inspire me. Find a circle of friends to study with. Find like-minded people like yourself and form a group for collaborative learning. The group can be of two people or more, but I would recommend not more than four. There will be questions like, what if he/she learns from me? What if he/she gets better and scores more marks than me? Keep these negative thoughts away. Your aim should be collective victory. You will experience joy when each one in your cohort emerges as a winner.

Peer learning is a win-win situation with many advantages:

1. Your doubts get expressed and cleared quickly, especially those that you might feel shy to ask your teacher.
2. You learn new techniques invented by your peers.
3. You learn from the mistakes made by your peers, without making them yourself.

4 Recreation comes as an added benefit. Studying will become fun. There will be jokes, laughs, and a bit of chitchat along with studying. But beware—too much gossiping will defeat the purpose of group study.

5 Most importantly, when you teach someone, you learn and retain a high level of information. Research shows that while people retain only 10% of what they read and 20% of what they see and hear, they retain nearly 90% of what they teach others (see Fig. 9.1). Teaching someone will help you to grasp the concepts better and will act as a revision for you, besides making you the favourite of the group!

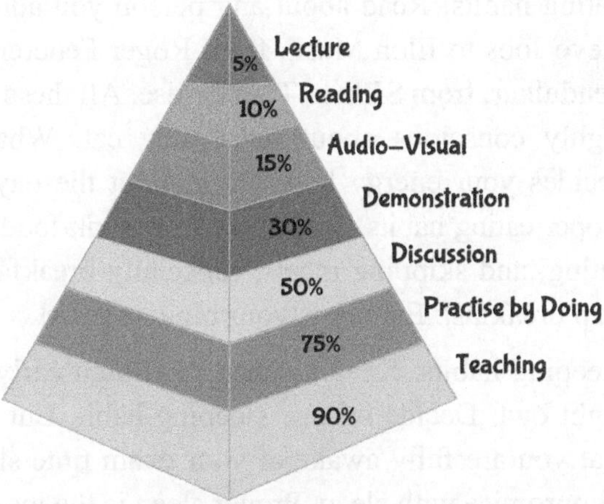

Figure 9.1 The learning pyramid

Note Remember, peer learning comes only after you do selfstudy. Protect your study time fiercely. Sometimes your friends will come to you for help during your self-study time; do not leave your self-study. At such a time, politely say "No" to them and fix some other time once you are done with your work. Then in the peer learning time slot, teach your friends whole heartedly. Happy teaching, happy learning!!!

Other Little Things

While climbing the pyramid, *Govinda* takes care of little things too. Things like where exactly to place the foot, at what pace to rise, etc. Similarly here are a few small things that will help you a lot!

1. Keep a pocket diary with you at all times. Write the to-do things. It is very effective. You can also write your daily homework there.
2. Also keep a small dictionary with you; you can use it for easy reference when you are stuck with language. This helps!
3. Eating habits! Read about any person you admire. From Steve Jobs to Elon Musk, from Roger Federer to Sachin Tendulkar, from SRK to Tom Cruise. All these people are highly conscious about what they eat. What you eat decides your energy level throughout the day. Develop proper eating habits. Stay away from junk food, late-night eating, and skipping meals, especially breakfast. Do not skip breakfast. Energy is your biggest asset!
4. Sleeping habits. Keep a schedule. Be an early riser or a night owl. Decide on one sleeping habit. But make sure that you are fully awake at your exam time slot. Do not compromise with sleep. Proper sleep is the most essential ingredient of concentration. Over-sleeping drains energy like nothing else. Know how much sleep your body requires. Generally, 6–8 hours of sleep is sufficient.

Concentration

If you ask *Govinda* what is his most prized quality? One that is more important than any other, you'll get the answer, CONCENTRATION!

The answer to distraction is concentration. To be more productive, improve your concentration. For amazing sure-shot results, improve your concentration!

> ***Concentration is the essence of all knowledge; nothing can be done without it.*** *—Swami Vivekananda*

A few tips on how to improve concentration. I find them helpful:

1 Start practising from classroom lectures. Start with small time durations of 15 minutes and be consciously focused in those 15 minutes. When you get comfortable concentrating for 15 minutes, increase the time duration gradually.
2 Keep your mobile away while doing homework.
3 Meditate, if you can, at least thrice a week.
4 Generate exam conditions at home and practice giving exams in such conditions.

Concentration is the other name of excellence. Embrace concentration as a life-enhancing habit. If you are playing, concentrate. Family time, concentrate. With friends, concentrate. Studying, concentrate! This is the sure way of a fulfilling life. Champions are not the ones who always win races—champions are the ones who get out there and attempt, and work harder the next time, and even harder the next time. 'Champion' is a state of mind. Champions are passionate. They spare no efforts. They compete to beat themselves as much if not more than they compete to beat others.

Be a CHAMPION!

Scan the QR code to listen to the author on this topic

10

Time Management and Planning

In one of my classes (the day was Saturday) during discussion of homework, I observed that very few students were participating in the discussion that day as compared to other days. Upon investigating, I found that many students had not attempted the homework. It was a bit of a surprise for me, as the students of this class were sincere and this homework was allotted a week's time to attempt. On asking for the reason, one of my students, Yash, said that it is getting difficult to manage the homework of three subjects as well as revision.

Most of the other students heartily nodded in agreement, as if somebody has expressed their biggest concern. Then Yash said "Sir, today is Saturday, the day when you tell us strategies for better performance. Please tell us on how to plan and manage time." Just to know whether the rest of the class was also interested in this topic, I asked "Do you people want to discuss this topic?" The class unanimously shouted, "Yes Sir!!!"

This strengthened my belief that this is a genuine concern for almost every student. This class was no exception. I started the discussion with a few questions:

How many students set daily goals and work accordingly?	70% raised hands
How many students make weekly goals?	30% raised hands
How many students make monthly goals?	10% raised hands

Goals/targets are important to keep us focused. They keep us active and help us make most of the available time. Take a pen and paper (or diary), write what you want/have to do (goals), and allocate time for it. It can be a month's goal, a week's goal, or a day's goal. If it is a month's goal, distribute it into weeks and days. If it is a week's goal, distribute it into days, keeping your priorities in mind.

Amit Sir, how to set goal/target?

Sir Goal should be MAST. The whole class had a hearty laugh. MAST stands for measurable, achievable, specific, and time-bound.

Measurable.—Goals should be measurable (how much/many) so that you can track the progress and stay motivated as you get closer to your goal. For example, "I will revise 5 chapters of maths for the next test" is a measurable goal, as you decided for *5 chapters*.

Achievable.—I will revise all the Qs of Class 11 maths in a day. Come on! The goals should be realistic and attainable but at the same time challenging to enhance your ability. Set goals that are 10-20% more than your current capacity and work hard to achieve them. As you are able to achieve them, make them more challenging.

Specific.—Goals should be specific to stay focused. They should have details. For example, "I will revise the G- *and Vg-marked Qs of Exercise 1 of modules* of 5 Chapters of Maths."

Time-bound.—The goals should have deadlines (completion date/time) to keep us moving and avoid waste of time. For example, "I will revise Exercises 1 of 5 chapters *by ... ... date.*"

Along with setting goals in one's mind, it is a worthy practice to write them (your pocket diary can be an apt place to write). Read them daily. They will keep you focused on the aim and away from distractions.

Isha Sir, I set goals and make plans but after two days the plans begin to be delayed and I give up. What should I do?

Sir First of all, understand that 'delay' does not mean 'defeat'. Do not get disheartened. Find out why your plan got delayed. Were you being overambitious and impractical or was there lack of effort or did something unexpected come up or something else? Take feedback, learn, and revise the plan. Second, when you make a plan, keep a bit of time as buffer time for unexpected things/events and making up for delay. It will help you keep up with the plan.

Rahul "Sir, I used to make a time-wise daily timetable for selfstudy as follows:

2:00 pm — 3:30 pm	Maths homework
4:00 pm — 5:30 pm	Physics homework
5:30 pm — 6:00 pm	Snacks
6:00 pm — 7:00 pm	Chemistry homework
7:15 pm — 8:45 pm	Revision

However, most of the time I am not able to follow it exactly. For instance, when snacks are tasty in our hostel mess, it usually gets late to come back to study. *Kabhi kabhi mann hi nahi karta padne ka* (sometimes, I do not feel like studying) and thus breaks between study slots get stretched for longer than allotted, and so on. A delay in one of the time slots disrupts the entire following schedule. As I am not able to follow the daily schedule, I stopped making it. How to make a daily schedule which can be followed?"

Sir It is human nature to get bored sometimes following the same routine. It is normal. Do not be too harsh on yourself. Cut yourself some slack. And yes, food should be enjoyed when it is tasty, as I guess mess food and tasty must be a rare combination in your mess also as is the case with most messes! Now, the schedule set up with strict timing is challenging to follow minute-to-minute as unplanned events may turn up. If it does not work for you, I would recommend making your timetable duration-wise in place of time-wise. For example, your timetable would look like

1.5 hour	Maths homework
1.5 hour	Physics homework
0.5 hour	Snacks
1.0 hour	Chemistry homework
1.5 hour	Revision

This way you will have a flexible schedule, scope for accommodating unplanned events, and you can interchange slots depending on your mood. However, take the caution that flexibility should not lead to laziness and procrastination. Set checkpoints of time to ensure effective execution of the timetable. For example, you may set a checkpoint as : By 6 pm you should have studied 3 hours. Try it out. Before going to bed every night, recap your day's events, see what went as per plan and what did not. Take feedback and incorporate it into next day's plan. If this duration-wise mechanism does not work for you, try something else. But do not stop. *Failing to plan is planning to fail.*

Aman Sir, how many hours should we study daily?

Sir As much as you can, along with taking care of yourself. Do not forget recreation, proper sleep, and keeping yourself healthy. Sleep for 6–8 hours depending on the demands of your body. Slogging at the cost of necessary sleep is not fruitful. The JEE process is not a 100-m sprint. It is a one-/two-year marathon, where continuity and long-term performance matters.

Aman How long should we study in one stretch?

Sir When you are practising taking a test, then in one stretch, a 3-hour sitting is needed to train your body and mind. However, in routine self-study, taking breaks is important. Keep one stretch of half an hour at least. The maximum limit can vary from person to person. Whenever possible, keep your mobile out of your sight while studying. The notifications, messages, and what not break the flow of study. After a stretch of studying, when you feel exhausted, get up and take a break.

Sahil Sir, how did you prepare for the JEE?

Sir In the year 2004, towards the end of my Class 10, my uncle bought me a gadget from USA. It had a digital clock, timer, and stopwatch. At that time, when there were no mobile phones at our home, this gadget was new and fascinating for me. This gadget was my companion for two years of JEE preparation. It was always on my study table. Initially I used it only as a clock. In Class 11, I thought I used to self-study for 6–7 hours, but the output was not satisfactory. Then one day, I thought of measuring my effective study time. I used the timer mode of the gadget. When I would study using a table and chair,

I would turn the timer on. When I would get up, I would pause the timer until I returned to study.

At the end of the day, I was surprised to see the timer showing only 4 hours 15 minutes. From that day onwards, I started using the timer mode for knowing my *effective* input. When I was measuring self-study time without a timer, I was leniently (and sometimes unknowingly!) including leisure and in-between breaks in my study time. That increased the count, but the reality was different. Now when I was aware of my *real* input, I worked towards increasing it. Along with quality of the input, quantity also matters when it comes to JEE preparation.

Sahil: Sir, how did you increase self-study input?

Sir Chits.

Class Chitssss!!! What is that?

Sir Chits - pieces of paper. I used to set a target of a certain number of self-study hours for a day and make that many paper slips (pieces). One slip for each hour. For instance, if on some day my target was 6 hours of self-study, I used to make 6 slips of paper and keep them in my pocket. The timer was set for 6 hours. It would be on during studying and paused otherwise. As soon as it showed the completion of one hour, I would tear up one chit. This used to give me a sense of accomplishment, and there was excitement to tear all the chits up as soon as possible.

As the chits were always in my pocket, and my pocket always with me, the leftover chits were a reminder to finish the task. This kept me moving and eliminated distractions. The process was fun and each torn chit gave me a feeling of achievement and increased my confidence. The last chit used to be the most difficult bit. My mind would be exhausted and something inside me would request the last chit to remain untorn. And yet I had the will and the heart would say:

"NO. TEAR IT. TEAR IT. There is only one more chit."

Finally, my heart would win. The work, however, had to be done by my mind. To keep the mind motivated, I used to tear one-fourth of that last chit when 15 minutes were completed. This would cheer up the mind and provide micro-motivation to keep moving. Then the next onefourth, and the next one-fourth, and finally the last onefourth. Huh!!! DONE… These chits helped me a lot in stretching my abilities. Find out whether they will do it for you!!!

Yash Sir, did you always study, study, study…?
(The class started giggling ☺☺).

Sir No. During routine study, I used to take breaks in between. I used to keep most of Saturday evenings (other than those before a test or exam) completely for leisure and fun. No academic books or discussion. Nothing. Only recreation. From 6 pm to 10 pm, I used to go out with friends/family for a movie, to eat out, or for a walk. The concentrated leisure time gave me a change of environment and recharged my batteries. The next day, i.e., on Sunday, I would be back to study with 10–12 chits in my pocket. Sunday was not a holiday for me. It was a day for homework, revision, and to clear a week's backlog, if any. Then I would follow it up the whole week with sincere study. And then Saturday evening would return again!!! For me it was Saturday evening, you may keep some other day or develop some other creative way for recreation. (Just then, the bell rang…)

Class Sir, today is Saturday. Its party time!!!

Scan the QR code to listen to the author on this topic

11

How Not to Give up

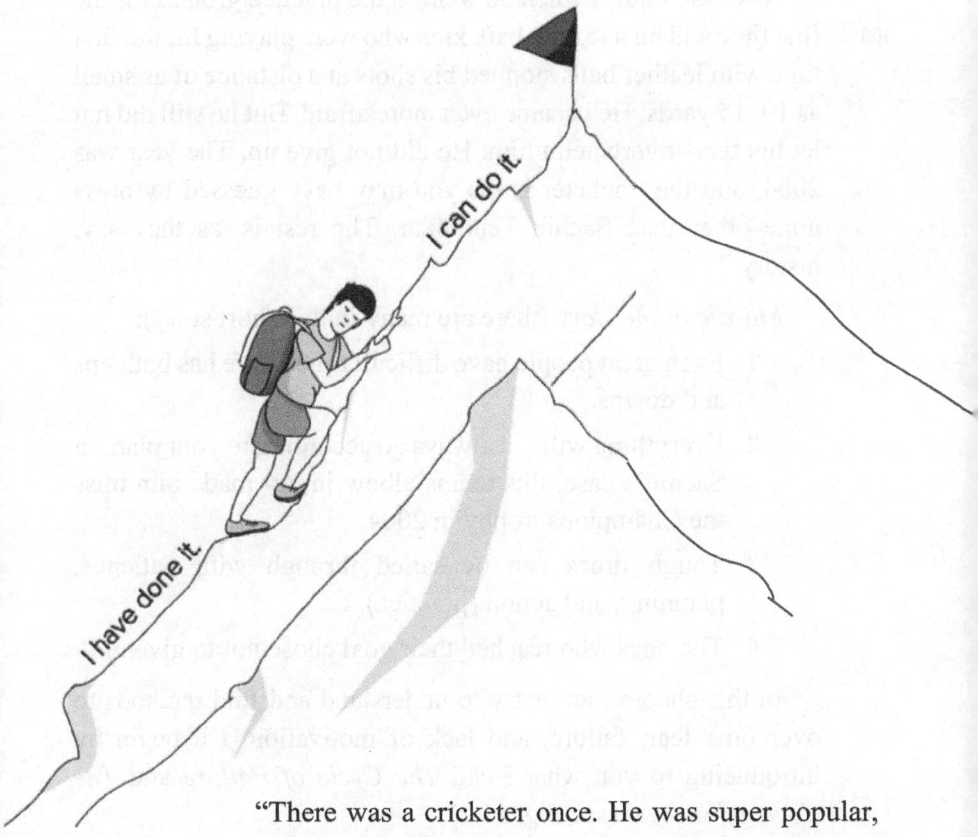

"There was a cricketer once. He was super popular, breaking all records one by one, carrying the hopes of a nation. Then in his prime, when everything was going his way, he got injured.

A career- hampering tennis elbow injury it was.
A man who played with one of the heaviest bats on tour could not lift his son's plastic bat! The injury led him to a stage where he was afraid that his career might well be over. Fear got into his head.
However, he did not give up. He patiently started recovering step by step.

After the injury, when he went to the practice ground for the first time and hit a leather ball, kids who were playing for the first time with leather balls stopped his shots at a distance of as small as 10–15 yards. He became even more afraid. But he still did not let his fears overwhelm him. He did not give up. The year was 2004, and the cricketer is (as you may have guessed by now) none other than Sachin Tendulkar. The rest is, as they say, history."

Morale of the story (there are many in this short story):

1. Even great people have difficult times. Life has both ups and downs.
2. Everything will not always go according to your plan. In Sachin's case, this tennis elbow injury made him miss the Champions trophy in 2004.
3. Tough times can be sailed through with patience, planning, and action (practice).
4. The ones who reached their goal chose not to give up!!

In this chapter, let us try to understand and find methods to overcome fear, failure, and lack of motivation. I'll begin by introducing to you what I call *The Cycle of Failure* and *The Cycle of Success* (see Fig. 11.1).

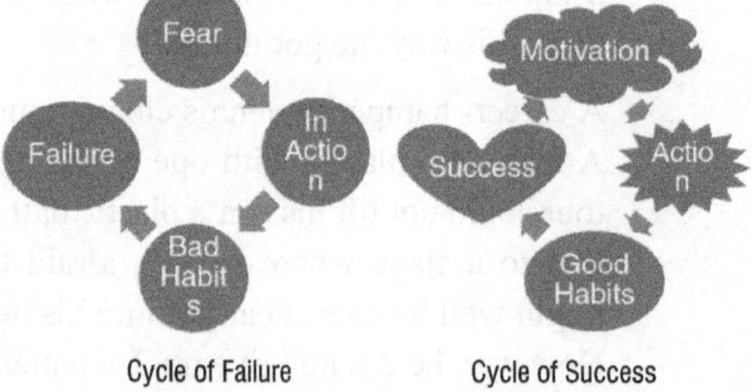

Figure 11.1 The Cycles of Failure and Success

Newton's law tells us that a body will continue being in its current state unless acted upon by a force. It is true in life too. Let us see how. Suppose that you are in a *cycle of failure* as well as in a *circle of fear*. Fear can lead to inaction, which then triggers an unresourceful state in our mind, keeping us from giving our best to a project.

This state becomes internalized and a negative pattern sets in. This then serves as a self-saboteur and ensures that we fail in the project. Our failure then becomes a self-fulfilling prophecy that we helped to set up. Similarly, the failure strengthens the fear we had, and the cycle repeats. Until a force is applied, the cycle will not break. Something has to be done. Force has to be applied.

The good news is, just like any successful person, YOU can also overcome this cycle of failure and convert it into the cycle of success. Each one of us has the potential to achieve our dreams—to break free from our own limitations and move towards becoming the best version of ourselves! But remember, just like it takes time to recover from injury, moving from failure to success also takes time as well as regular practice. To break this cycle of failure, we have to overcome our fear and plunge into action.

Overcoming Fear

> ***I learned that courage was not the absence of fear, but the triumph over it. The brave man is not he who does not feel afraid, but he who conquers that fear.***
> *—Nelson Mandela*

Fear is an unpleasant emotion caused by the threat of danger, pain, or harm. This is the Google definition of fear. What I want you to notice is, just like happiness and pain, fear too is an emotion. That means it is conquerable and you CAN conquer it.

Let us first contemplate the types of fear candidates generally encounter:

1. Short-term fear: Fear caused by actual events, such as
   (a) Bad performance in practice tests
   (b) Illness
   (c) Comparison with others
2. Long-term fear: Fear caused by imaginary events, such as
   (a) Fear of the future

Coping up with bad performance in tests

After an exam, which does not turn out to be as we expected, we feel depressed and embarrassed. To add to the embarrassment, we are asked needling and mocking Qs like "How did the test go?"

We start doubting our self, lose hope, quickly come to negative conclusions, and so on... But wait! Before declaring yourself unworthy, have you analysed what went wrong? Yes, I am talking of test analysis. I know it is very difficult to face that demonic test paper again, but that is necessary. If you think with a cool head, the test paper is actually a responder, who is ready with feedback for you, which you can choose to receive or ignore.

There are no failures, only feedback.

Upon analysis, generally you will find three reasons that led to bad performance in the test:

(a) Entering into panic mode in a high-pressure situation. You will find that of all the Qs answered wrongly in the test, you can easily solve many of them at home.
(b) Silly mistakes during the test.
(c) Poor preparation for the test.

Remedy Analyse your test and be flexible and open to receiving feedback whole heartedly, honestly, and positively. Take responsibility for your failure and accept the reasons. Learn from it. Do not blame others. If you identify 'panic' or 'making silly mistakes' as the reason for bad performance, practise taking tests at home, once a week, to get habituated to the test environment and thereby reduce silly mistakes. Refer to Chapter 6 for more on improving test performance.

If you find that 'poor preparation' is the culprit, the solution is simple—prepare well the next time! Give preparation your best shot. Understand that humans make mistakes. You also did, but let us not repeat them. The test that is over is over in the past and cannot return.

Whatever has happened has happened. If you want to cry about it, cry for exactly 300 seconds, criticize yourself and your past, but not a single second more. Not even a single millisecond more. Get up, splash cold water on your face, eat something, and get back to work. It is action time!!!

Coping up with Illness

In student life, becoming ill is not uncommon, especially when you are away from your home. Illness might mean loss of time and classes. It may stress you out and preparation may get adversely affected. In this situation, we are faced with certain choices. We could choose to take the stress and weep for our misfortune or bravely face the situation and plan how to control the damage. By being both ill and stressed, we do nothing good but add an extra burden on the body, making it even more difficult for it to recover.

Take the second path; it is fruitful—choose NOT to be stressed. I have seen many students facing illness bravely. After they recover, they patiently cover the part they missed while being sick, and in the long run they achieve their goal.

Sometimes you might have to take practice tests in your coaching centre/school during an illness. Do not lose heart. Be brave!! I know people who were sick on their JEE day, *yes their JEE day,* and they managed an AIR under 500. On that day, their experience of taking practice tests under adverse conditions played its role. The determination and 'walking through' attitude led to the glory.

One such example is Amit Malav. Amit, in the middle of his Class 12, was diagnosed with a rare disorder. He had to be hospitalized. He could not sit for long hours and most of the times had to lie down on the bed.

However, his spirits were high. He did not let go of his goal even in this condition. For him to continue his studies, his father would hold books in front of him sometimes for more than 10 hours a day. He gave exams under intense medication to overcome pain. His determination defeated his illness and he secured All India Rank 87 in JEE 2013. His highly inspirational journey is covered in detail in Chapter 12 (*Those Who Triumphed*).

Remedy In the first place, take preventive measures to avoid falling sick. Eat well, sleep well, and do exercise, yoga, and meditation. Even after these measures, if you fall sick, take rest and follow the path of the warrior. Be brave, make a new action plan to cover lost ground, and act on it!!!

You have not lost time, because time cannot be lost. You simply have to use the remaining days wisely. Taking this path is a way of telling destiny "Kuch bhi ho jaye, hum darne vale nahi hain" (Whatever happens, I will not be afraid).

Coping up with Comparison with Others

"Seekho Rahul se kuch. Uske 98% aaye hai, aur tumhare?" Does this comparison-coated reprimand sound familiar to you?

Some parents/teachers use this approach in good spirit for the betterment of their child, but it does not produce the expected result always. The mentality of comparison unknowingly gets coded in the child's brain, which repeatedly haunts him/her: "What if my friend performs better than me?"

It is the most futile kind of fear. Each person is unique. Each person has his/her own speed and way of acquiring knowledge. To compare and assess our self-worth without the attitude of learning is a waste of our energy. It shifts our focus from our self to someone else.

Remedy If you find someone admirable or find yourself being jealous of someone, ask yourself—what quality/strategy does s/he have that I like. Can I replicate it in my life? If you can copy it and it works for your individual style, then go ahead. If not, discard. The quality/strategy works for that individual. You will have to find your own strategy.

Learn from others and compare/compete with yourself! You cannot control the actions of others, so you cannot control their results too. What you can control are your actions. Simple choice, right? Tell yourself every day that you will be better than how you were yesterday! Focus on making your results better than the last time. Better, better, better... and you will tend to become *the best of you.*

Fear of Future

What will happen in the next test? What if I am unable to revise all the chapters in time? Will I get selected or not? What would people say about me if I do not get the desired rank? ... The list is too long to be listed here.

Nothing in life is more amusing than the unnecessary anxiety that we endure, and generally create ourselves. We suffer today with anxiety for something that might never happen. Or worse, we waste time and stop preparing, thus creating that future ourselves!

> ***Remedy : Do not dwell in the past, do not dream of the future, concentrate the mind on the present moment.***
>
> —*Buddha*

Stop living in an imaginary world. Today is all you have to prepare for tomorrow. Make a plan and swing it into action! Put your best foot forward!! The best way to predict the future is to create it. Fear leads to inaction, to excuses. Take responsibility. Become a person who acts, not one who fears and withdraws!

Call to Action

> ***The moment you fear, you are nobody. It is fear that is the great cause of misery in the world. It is fear that is the greatest of all superstitions. It is fear that is the cause of our woes, and it is fearlessness that brings heaven even in a moment.***
>
> —*Swami Vivekananda*

No more fear now. You are a brave person. Let us move towards fearlessness by applying the techniques mentioned above to tackle short- and long-term fears.

The bridge between a dream and reality is action. Set your target, make a well-formed plan to achieve it, and action it. Planning without implementation is futile. Execution of a plan through action is the most important part of your journey. In this journey, mistakes may be made. Some failures may have to be faced. It happens. Do not get disheartened. Mistakes are proof that you are making the effort.

Understand this—every failure brings with it an opportunity to learn from mistakes and to not repeat it in the final exam. That is it!! Do not let failures stop your actions. If you did not achieve success following one way, strategize, and find some other way. BUT do not stop till you succeed.

To succeed, you have to make sacrifices. There is no doubt about it. Many a times you will have to say "No" to friends and movies. If you are living away from home for studies, you will have to sacrifice the best food, fun at celebrations, weddings, and leave behind all the pampering, and come out of your comfort zone. There is only one goal to achieve, everything else has to wait.

> ***To succeed in your mission, you must have singleminded devotion to your goal.***
>
> *—Dr A.P.J. Abdul Kalam*

There will be times when you will be demotivated. It happens with everyone. Yes, even with people who have finally achieved AIR-1. When such a moment comes, read the goal that you have kept in writing (refer to Chapter 9).

Read something motivational. I would recommend sticking some inspirational posters or quotes (or your goal) in your study room. They are a source of inspiration and motivation.

> ***I hated every minute of training, but I said, "Don't quit. Suffer now and live the rest of your life as a champion."***
>
> *—Muhammad Ali (the Greatest Boxer)*

Success

"Boond boond se ghada bharta hai" (Drop by drop a pitcher fills). This applies to the process of improvement also. Anyone who thinks of becoming the best overnight is setting himself/herself up for failure for sure. And this expectation of getting overnight success is the main culprit that disheartens many candidates.

Suppose you score 40% marks in a test, and then you expect to jump to 90% in the very next test. This is being impractical. If it happens, then that is wonderful but if not and you only improved to 45%, then please do not lose your morale. You improved. It is success.

You got better. In the next test, try for better and when you achieve it, push the bar up. Keep improving. Give success time to grow into a Banyan tree. Keep becoming a better version of yourself. When dreams are nurtured with hard work, dedication, determination, and smartness, the success is bound to be magnificent.

Success is not only associated with the end result. The qualities that you are acquiring in the journey are themselves a lifetime treasure. Yes! The fighting spirit that is being kindled inside you, the habit of working hard, cultivating the 'will not give up' attitude, and enduring all the hardships are a priceless collection of qualities that will serve you all your life. Do not make the mistake of undervaluing them.

These qualities will form the foundation of your future endeavours. Stand up and seize the moment. Do not leave a single stone unturned. Walk through. Now!!!

All the power is within You; You can do anything and everything. Believe in that. Do not believe that you are weak. Stand up and express the divinity within you.

—Swami Vivekananda

Scan the QR code to listen to the author on this topic

12

Those Who Triumphed

Achin Bansal

(AIR-1, IIT-JEE 2007)

Achin was brought up in Kotkapura, a small city in Punjab. His father is a doctor and his mother is a homemaker. In his schooling days, not much was known about competitive exams such as NTSE (National Talent Search Examination), KVPY (*Kishore Vaigyanik Protsahan Yojana*), or any Olympiad in his school in Kotkapura. So till Class 10, he did not give any of these exams and focused only on his school curriculum. He topped the school in Class X Board exams.

After Class X, he had to pick a career stream. "Although my father and my uncles and aunts in the family are doctors and we have a medical setup in Kotkapura, I had the freedom from day one to choose whatever profession I wanted to go for," says Achin.

He liked Maths, so he chose engineering. It was at the end of Class 10 when he came to know about his father's friend's son who was in Kota for JEE preparation. After talking to him, Achin decided to go for JEE coaching in Kota. Achin was to go on a Europe trip with his family before the coaching classes would start, but due to some issue with Visa, it got postponed. "It was going to be my first foreign trip and I did not want to miss it," Achin recalls fondly.

After the Europe trip, he came to Kota 15 days after the classes had started. He was allotted the 20th batch from the top. The first week was very tough. He was away from home and missing his family. He was not eating well, as the food did not taste like home. He had to set his room and other things in order.

He was also having a difficult time understanding the things taught in class, as he had missed the initial classes. He became weak and once he fainted in his room! Alarmed, his grandfather came to Kota. He took care of Achin and helped making his life functional. Achin made some friends, who helped him cover up the lost classes. The first test at the coaching centre was after a couple of weeks.

> ***"In those weeks, all I did was study, eat, sleep, study, eat, sleep...,"*** —Achin reminisces.

In the first test, he got the 58th rank in the coaching centre. In his coaching classes at Kota, ranks were given to all the candidates from all the batches of Class 11. So it was a rank among more than 2500 candidates. It was a big success for him. His confidence received a boost. From the 20th batch, he got promoted to the top batch.

> ***"Being in the top batch had some challenges associated with it. The pace of teaching was way faster than it was in the earlier batch. Expectations were more. The environment was completely different for me,"*** —Achin says.

He found it difficult to keep up with the pace, and his rank fell from 58 to around 130 in the second test. He got demoted to the second batch. However, he did not let the incident demotivate him. He took this as a challenge and worked hard systematically. In the upcoming tests, his rank improved and after a couple of months, when shuffling of batches happened, he made it to the top batch again.

He was doing well, got under Rank 20 sometimes, but could not grab a rank in the top 10 until, in the last test of Class 11, he reached Rank 3. By this time his cumulative rank was 13. His journey in Class XI started from Rank 58 and rose to 13, going through ups and downs.

After Class XI was over, there was a month's break before starting Class XII. Achin went home in this break. "I was so satisfied with my performance in the last test that I did not study at all in the break. I got overconfident," Achin shared. After the coaching classes resumed, in the first test of Class 12, where the major syllabus was from Class 11, his test rank crashed to 80. He barely maintained the first batch.

> ***"It was an eye opener for me. I realized that I had to be consistent in my preparation and change my attitude towards learning. What if such a thing happens in the JEE?"* he says.**

Achin got focused, rectified his mistake, and made a comeback, getting a test rank under 10 in the second test. He maintained his test rank under 10 in most of the Class 12 tests with the best of Rank 2 in one of the tests. His wish of getting Rank 1 in the coaching classes remained unfulfilled. Who knew what fate had in store for him!!

In December, the syllabus of Class 12 got completed at the coaching centre. The JEE was nearly three months away, and the Boards were on the way. In the year 2007, the JEE was held on 8

April and the AIEEE (All India Engineering Entrance Examination, which is now JEE Mains) was held on 29 April.

> ***"The last three months before JEE are very crucial. I would give credit to the systematic work and planning that led to efficient revision during the last months,"***
> —Achin shared.

In the two-year preparation, Achin had made organized solutions of the questions of the DPPs (Daily Practice Problems) and Sheets (topic-wise collection of Qs). He used to solve the DPPs on loose pages and attach the solution pages with the DPPs.

> ***"The solutions gave me the confidence and peace of mind that I could visit them anytime and more importantly that I had solved the questions. The solutions prevented waste of time and avoided a panic situation in case I was not able to solve any question,"*** —Achin said.

While revising the DPPs and Sheets, he did not solve each and every question. For some of the questions, he would read them, think of (for some of them, write down) the concepts and equations applicable, and verify from the existing solutions whether his concepts/equations were correct.

> ***"This way I ensured that I knew how to solve the problems, and saved time in calculations. For calculation practice, I used to solve some of the typical questions till the end."***

Systematic notes with important things marked in them helped him cover the large content in a short time. After the Boards got over, Achin started to train his body and mind for the JEE.

"For the last 15 days before JEE, I used to wake up at 6 am, have breakfast, and study without any break from 9 am to 12 pm. Then I would take lunch, and again study or take a test from 2 pm to 5 pm using a table and chair, as if sitting in the final JEE exam."

One day before the JEE, Achin just skimmed through some of his notes with a relaxed mind. He had the feeling of satisfaction.

"Whatever I could have done had been done to its best, and I was satisfied with my preparation."

This feeling comes through honest preparation. It helped him keep a cool head before the exam. With this feeling, he went for the JEE. In the JEE, he attempted maths first (which was his strongest subject). Alas! He could not solve the very first question of his strongest subject in the first paper. This is where his training and test strategy saved him. He left the first question after spending 2–3 minutes on it and moved on to the second question. He could do the second problem, the third, the fourth, and so on. When he was able to do most of the questions in maths, his confidence went up, and when he revisited the first question he could solve it!! Achin scored 152/162 in maths!!

"My test strategy was to let go of difficult questions in the first attempt and not spend more than 45 minutes on each subject in round one. Luckily I could attempt all the questions of maths in the first paper in 45 minutes, as maths was easy that year."

After the papers were over, Achin calculated his marks using the answer keys provided by different coaching classes and was confident that he would get an AIR under 15 or 20. When the results came and he logged in with his roll number, he saw AIR-1 on the computer screen. His first reaction was "there might be an error." A digit might be missing after 1. So he checked the result again and again! His JEE total score was 429/486. His dream came true. The hard and systematic work done had rewarded him. He was AIR-1 in IIT-JEE 2007.

AIEEE OMR ka locha.—After giving the JEE exam (he knew he would crack it), Achin took the AIEEE for fun, which was three weeks after the JEE. The OMR of AIEEE had to be filled by pen. By error, he filled the answer of one of the questions into the answer of the next question, and subsequently filled the next 7–8 answers at wrong places, till he realized his blunder.

Nothing could be done, as the answers had been filled by pen and could not be erased. But as people say, *fortune favours the brave,* he got three answers correct from those 7–8 questions. He bagged AIR-47 in AIEEE.

Hostel life.—In those two years of preparation, Achin made friends, with whom he used to discuss a lot. These were people who were taught by Achin after he would complete his self-study.

> ***"I was a shy person, who did not ask doubts in class. For me, discussion with friends was the doubt-solving mechanism. Teaching someone is a win-win situation. For me, it acted as a revision. Also, while teaching my friends, I realized that there were some gaps in my own learning, which got discovered when my friends asked doubts about them. Sparing time for discussions after my self-study was very helpful,"***
>
> —Achin recollects.

He used to have fun with his friends as well. After each test, he would go out with friends or play games such as Ludo and Snakes and Ladder. When Achin's grandfather came to his hostel, he gave Achin a copy of the *Hanuman Chalisa*, which he used to read daily.

It acted as a source of motivation for him. Achin did his graduation from IIT Bombay in the Computer Science and Engineering branch. He is at presently working in New York, USA, in a financial services firm, Morgan Stanley.

Sai Sandeep Reddy (AIR-1, IIT-JEE 2013)

Palleria Sai Sandeep Reddy, also known as Sai, was the AIR-1 holder of IIT-JEE 2013. Sai is from Andhra Pradesh. His father eaches in a school and his mother is a homemaker. After completing his primary education in a village, Sai moved to a town in Class 7. He was a mathematics lover from childhood.

His journey started with winning a town-level maths competition in Class 7. In Class 8, he won a country-level maths competition conducted by AMTI (The Association of Mathematics Teachers of India). This boosted his confidence and he started preparing for the IMO (International Mathematics Olympiad).

In Class 10, after clearing the first two (regional and national) stages of IMO, Sai, being among the top 35 candidates of India, was invited for a month-long IMO Training Camp (IMOTC) in Mumbai.

Sai had a keen interest in maths, so there were two options in front of him: research in maths or prepare for the JEE. After talking to friends, relatives, and teachers, he took the decision to prepare for the JEE. In Class 11, he joined a known coaching institute in Hyderabad for JEE preparation where many candidates were preparing from Class 9.

> ***"Initially it was very difficult, competing with my fellow students preparing from Class 9. Only when the maths was tough, I was able to beat them in the tests. But as time passed, I was able to give them a tough fight, and thereafter I consistently scored well,"*** **says Sai.**

 For Sai, mathematics came naturally, but remembering chemistry was difficult. So he used to devote nearly half of his preparation time to chemistry and the remaining half was distributed equally between maths and physics.

> ***"In maths, I focused on maximizing the learning from a given number of questions rather than going after solving as many problems as I could. In physics, I used to go into the depth of the concepts, understand them fully, and then solve the questions. Just applying the formulae without understanding the concepts is not of much help. For chemistry, the mechanisms are very important along with remembering things." The NCERT books are a must read for chemistry. I also used to make short notes having all the important points and formulae of the three subjects,"***
>
> —recalls Sai.

His day used to start at 6:30 am. Every day he used to attend lectures of 2 hours each for the three subjects. Then, after a lunch break, he used to do his homework. Sometimes they had sessions of 2 hours in the presence of a teacher, where they were given questions for practice and then they used to discuss them in groups. Studying for around 12–13 hours, Sai's day would end at 10 pm.

By July of Class 12, he had completed the entire JEE syllabus for the first time and thus had nearly a whole year for revision.

> ***"First of all, I noted down my weak chapters and worked on them. Thereafter, I used to revise around three chapters per week. From December onwards, I focused on solving previous years' JEE papers, other institute papers, and revising theoretical aspects, especially of chemistry. I had to revise chemistry every now and then to keep it in memory."***

Every Sunday, he used to take tests at his coaching centre. He experimented with many test strategies and then settled for one. "The strategy that suited me the most was starting with chemistry. I used to do chemistry for around 40 minutes, then physics (1 hour), and then maths (1 hour); then, the last 10–15 minutes for whatever was left. The most important thing that I took care of was not to get stuck with the difficult questions initially, as all the questions were of nearly the same marks," tells Sai.

Sai scored 98% in Class 12 from Andhra Pradesh Board. When I asked Sai how he managed such an awesome Boards score along with the JEE, he modestly said, "I had Andhra Pradesh Boards, which I think is easier as compared to other Boards. PCM was already getting prepared for along with the JEE preparation, and the other subjects were not very difficult to manage." After the Board exams were over, Sai made final preparations for the JEE Mains. As Sai was mainly focused on JEE Advance, he gave more importance to the topics that were in the syllabus of JEE Advance as compared to the exclusive topics of JEE Mains.

His JEE Mains test centre was in a different city, where he arrived late in the night before the exam day. To add to the improper sleep that he had the night before, he became a little sick. But his JEE Mains went well and he bagged AIR-80 in JEE Mains with a score of 311/360.

Other than JEE Mains, he gave Andhra's EMACET (Engineering, Agriculture and Medical Common Entrance Test) in which he topped, BITSAT (scoring 418/450), and a few others. Sai also prepared for IMO in Class 11 along with JEE preparation. He cleared the first two stages and again made it to the IMOTC, being among the top 35 candidates. He was ranked 7th in the training camp, but only the top six candidates got a chance to represent India at the international level!

When I asked him, "Did you expect AIR-1 in IIT-JEE?" he fondly said, "No. I was confident about getting ranked in the top 50, but AIR-1 came as a pleasant surprise."

Tips from Sai to Future JEE Aspirants

1. Enjoy the preparation process. Perseverance will come automatically.
2. Time management and planning are important. Divide your time among the three subjects according to your strengths and weaknesses, and study with a plan.
3. Interact with your teachers as much as you can
4. Take breaks at regular intervals. Avoid continuous slogging.

Sai completed his B.Tech from IIT Bombay, Computer Science Department, in 2017. Further, he is planning to pursue a PhD in Computer Science from CMU (Carnegie Mellon University). Sai wishes all the very best to future JEE aspirants.

Palak Jain (AIR-28, IIT-JEE 2013)

Palak is from Udaipur, Rajasthan. She is from a family of doctors. Her parents and her sister are all doctors. But Palak had an interest in mathematics. In Class 9, she joined a coaching class in Udaipur to prepare for competitive exams such as NTSE, NSO (National Science Olympiad), and IMO (International Mathematics Olympiad). Her coaching continued in Class 10.

In these two years, she cleared NTSE and was awarded a scholarship. She also cracked NSTSE (National Science Talent Search Examination) with AIR-5. To add on, she qualified IMO with Rank 11.

> ***"My favourite subject was maths. Preparing for competitive exams, my decision to go for the JEE got gradually firmed up from Class 9, and in the middle of Class 10, I decided to go for it,"*** —recalls Palak.

In Class 11, she decided to prepare for the JEE with the same institute in her hometown. Palak shared, "At home I had the support of my parents. At my coaching centre, the teachers were really good, and as we were a small number of candidates as compared to big coaching centres, we got personal attention too. Competition among the candidates was healthy and sufficient to keep striving for better performance. I did not feel the need to go to any other city."

During preparation for the JEE, she used to self-study for around 8–9 hours. *Marking and test analysis* played a very important role in her performance. Palak used to mark the tricky and good questions of the DPPs and Sheets that went into the depth of concepts. She did this for every subject. She also used to mark important sections in reference books for revision. After each test, other than evaluating her score, she used to redo all the questions that went wrong in the test and do a detailed test analysis.

> ***"I used to note down all my mistakes made in the test and the cause behind each of them. Then, after segregating the mistakes and analysis, came out the areas where I was going wrong."***

Once she knew the real culprit, she started improvising. Elaborating on this, Palak says "We all know that we make silly mistakes in tests, but what gets missed is the cause behind them. What is really important is to go after the cause of the mistake. This is the first step towards improvement."

In Class XI, Palak got selected in KVPY and was awarded a scholarship for the same. In Class 12, she appeared for NSEP (National Standard Examination of Physics). However, not everything went as she wished. She did not qualify for the second round.

"The NSEP result set me thinking and doubting myself, as most of the top JEE rankers clear the first round. I thought, even if I am in the top 3 in my coaching centre, would I be able to perform well in exams like JEE since the NSEP did not go well?"

However, Palak did not allow this result to depress her. She took it as feedback. At home she did a detailed analysis of the NSEP paper and spotted where she went wrong. She talked to her physics teacher on how to improve.

"The NSEP result turned out to be a good thing for me. I started focusing more on physics and worked on my weak areas. That also taught me not to jump to conclusions based on temporary failures or successes. Not doing well in one or two tests does not at all mean that you do not deserve anything,"

—Palak shared.

To prepare for the Board exams, Palak went through the NCERT books along with the JEE preparation in Class 12. As a particular chapter was completed at her coaching centre, she would read the chapter from the NCERT textbook. "Studying from NCERT books throughout the year did not make them a burden for the end.

While reading the NCERTs, I used to mark the important sections in them from the Boards perspective," recollects Palak. Her NCERT markings saved her a lot of time during the Board exams. This helped her continue studying for the JEE Mains while preparing for the Boards (even during the Board exams).

To practise writing subjective answers for the Boards, she followed some standard CBSE guide books having sample questions and answers along with the NCERT books.

For the JEE Mains preparation, she used to practice taking tests for being in flow with the objective pattern. The JEE Mains was less than two weeks after the Board exams. Her markings and systematic work helped her achieve both the milestones. Palak fetched 94% in the Board exams and her JEE Mains score was an outstanding 340 out of 360.

After JEE Mains, the preparation for JEE Advance began. First of all, she noted down her weak topics and started working on them. She had a simple strategy: revise and strengthen things that were already taught and go through the test analysis of previous tests. She did not refer to any new book or material in the last month other than new questions in the tests that she used to take for practice.

> ***"In some of the topics, I realized that I had forgotten things. This happened especially in chemistry. So I revised the notes and questions I had marked."***

While preparing, she tried many different test strategies. Taking learnings and feedback from them, she decided to adopt the test strategy of solving the paper in two rounds. In the first round, she used to attempt easy questions and mark the tough ones for the second round. She used to decide the time that she would give in the first round to each of the three subjects. Then, once the first round was completed, she used to visit the marked tough questions in the second round.

> ***"I used to leave questions that were not getting solved in the first round and not get emotionally attached to them, then come back to them in the second round. Also, when the last minutes were left at the end of the test, and there were some questions left or marked for revisit, my strategy was to focus only on a few of them and not jump between questions in panic."***

Before the JEE exam, she perfected this strategy by practising it in her coaching tests and the tests she took at home. One day before the JEE Advance exam, she did not study much.

She had not left any doubt to be cleared for that day. She just talked with her parents and *chill mara* (relaxed). Her hard work, dedication, strategies, markings, and analysis led her to AIR-28 in IIT-JEE 2013.

She opted for IIT Bombay, Computer Science branch. She had been a part of the IITB Table Tennis League and Convener of *Samwad*, a social club at IITB. In the first-year holidays she completed a 26-day mountaineering course conducted by the mountaineering group at IITB, and the list goes on… She did her internship at Google, San Francisco, and has got a placement at Google, India.

Amit Malav

(AIR-87, IIT-JEE 2013)

Amit was brought up in a small village of Rajasthan named Takarwara, which is around 50 km from Kota. His father is a farmer who had completed his BA (Bachelor of Arts) and joined the family farming occupation. In Takarwara village, there was only one school whose condition was pathetic. To improve the quality of education, Amit's father opened his own school. Amit got his primary education in it and came in the merit list in the Class 8 Board exams.

In Class 9, Amit joined the school in the nearby town and completed Classes 9 and 10 from there. There were times when the school did not have teachers for some of the subjects. He studied those subjects on his own. In Class 10 also, Amit secured merit in RBSE (Rajasthan Board of Secondary Education) Board exams.

During Class 9, he developed a pain in his joints, which continued in Class 10 also, but nothing could be diagnosed. In Class 11, he decided to prepare for JEE and came to Kota. As he had studied in Hindi medium till Class 10, he joined a Hindi medium batch in one of the known coaching institutes of Kota.

In the first two tests of the coaching centre, he scored very well. His performance got noticed. His teachers suggested him to shift to the topmost batch to study with the best brains of the institute. But the problem was that the top batch was an English medium batch. Amit took a bold step and moved to that batch.

> ***"Initially I had to work on the language part along with the subject. The teachers were very supportive. They used to translate typical English words into Hindi. Also, I was provided a person who would translate the English study material into Hindi for me. This helped me a lot to settle down into the new batch,"* says Amit.**

Amit enjoyed problem-solving a lot. His favourite subject was maths. He studied sincerely. He had kept a simple rule for himself: "Not to pile up the stuff. Finish the work at hand." He used to revisit the content taught in class and complete his homework on the same day at home. Timely completion of work led to a strong academic foundation without unnecessary pressure.

For revising a chapter, he used to directly start with solving the hard problems of that chapter. This way he used to gauge his level of understanding of the concepts. His test strategy was to start the test with the strongest subject (which was maths for him). This would build up his confidence during the test. After the test, he used to analyse and identify the weak areas to work upon.

In one of the tests, he got all questions correct in physics except a question from the chapter on rotational dynamics. This somewhat upset his physics teacher, which Amit took to his heart. He re-solved the Sheet of rotational dynamics *30 times* as well as solved all the related problems from the book I.E. Irodov!!!

During the whole year of Class 11, the joint pain never left him. Sometimes it was unbearable. He struggled with the pain but did not stop attending classes and kept up with his studies. He was helped by the Director of the coaching institute, who connected him to renowned doctors. The doctors kept him under medication for the whole year. It was at the end of Class 11 that Amit was diagnosed with a rare rheumatic disorder.

He bravely accepted the situation and continued his JEE journey in Class 12. In July 2012, two months after the beginning of Class 12, the situation worsened—the pain aggravated. He had to be taken to Jaipur for consultation and then to AIIMS in Delhi. He was hospitalized for 15–20 days. It was a stressful time, as a lot of time, energy, and money were spent in the treatment.

After hospitalization, the treatment had to be continued in his hostel room. He was in no condition to attend classes, as he could barely sit continuously for half an hour. Most of the times he had to lie on the bed. From September onwards, Amit could not attend any of the coaching classes.

The pain broke his body but could not break his spirits. He requested his teachers to teach him at his room, which they accepted whole-heartedly. He used to study lying down on the bed. His coaching institute did not leave any stone unturned to facilitate his preparation.

"I was blessed to have support beyond words from the institute. My teachers taught me and provided mentorship and moral support. Also, I cannot thank my hostel caretaker enough, who went the extra mile to take care of me," **says Amit.**

His father, being a farmer, had a tough time bearing the cost of his treatment. One dose of a particular medicine that Amit had to be given cost around INR 82,000, and he was to be given 3–4 such doses. At such an adverse time, his coaching institute came forward to bear the expenses.

Amit did not give up. He continued his JEE preparation lying down on the bed. To help him in reading, his father used to hold his books in front of him for long durations.

The most challenging subject to study was mathematics, as it has to be practised by writing and there were times when Amit found it difficult to hold a pen, leave aside sitting. "Maths was a challenge, but the good thing was that it was my strongest subject. Gradually it got managed. Also, I was fortunate enough that I had a strong command over the content of Class 11. Its revision did not take much time during Class 12."

The heavy doses of steroids and antibiotics had their side effects. Nearly for the whole year of Class 12, he had rheumatic fever due to the medication. Fever along with the pain made things worse for him. However, Amit had a strong motivation, which led him to sail through the tough waters.

"I had to prove to the world and more to myself. My parents had faced hardships educating their children. I had to make their dreams come true. Also, in my condition, I could not do much lying in the room, so I studied. My weakness somehow got turned into my strength!" **says Amit.**

Before his Class 12 Board exams, doctors advised Amit not to appear for the exams, as they were not sure if he could sit for three continuous hours. But Amit was determined to not let anything come in his way. Before each of the board exams, he took pain killer injections, which helped him sit through the exams.

The Boards went. The JEE Mains came. His condition was the same. He appeared for the JEE Mains with preparation, determination, and injections. The JEE Mains results came. He scored 325 out of 360 marks.

In Class 12 Board exams (RBSE), he had accomplished 91% marks. With these marvellous scores, he grabbed AIR-21 in JEE Mains. The intense medication that he had been going through had stressed his body. In the last 10 days before the JEE Advance exam, he did not study much and gave some rest to his body.

> ***"In the last 10 days, I solved only 10–20 maths problems per day. The rest of the time, I used to watch IPL cricket matches, some TV series, and relax. My friends used to say "Padh le bhai" (Study dude!) but I did not. I believe that for the JEE, what matters is the preparation of two years, motivation, and a relaxed mind in the last days."***

Before the JEE Advance exam, Amit had to take pain killers to ease out the pain. The chairs at the examination centre were not suitable for him, so he requested to use a pillow, which he was allowed after seeing his medical reports. The three hours flew by in the first paper. Then, the second paper. The JEE Advance papers had questions in both English and Hindi media, so he used to mark things in Hindi and then solve the questions.

When the results were declared, he had brought glory for himself and his family. Amit grabbed AIR-87 in JEE Advance.

His positive and not giving up attitude walked him through and led him to achieve his goal. Amit completed his graduation from IIT Bombay, Computer Science Department, and has secured a placement in a technology firm in Bengaluru.

Scan the QR code to listen to the author on this topic

13

The Last Month

Are you all set for the last month of preparation?

Yes, the pressure is immense, expectations are high, but do not worry, you have the support of the hard work that you have done. Have faith in yourself—you have worked hard and strategically. You have done a great job till now. It is the time to actualize the hard work you have put in. This is the time to manifest your destiny.

Use this last month productively and win your prize. Say with me loudly "I can do it. I will do it." Now let us prepare systematically and, again, strategically.

The Period After Board Exams

If you are in Class 12, then you have possibly prepared for your Board exams also. After the Board exams (which generally end in the last week of March), depending upon which Board you have, you will get a gap of around 7 to 15 days after your Board exams till the JEE Mains. So, in this small gap, how can we revise the vast JEE syllabus?

The answer lies in taking help of the markings we have made (as explained in Chapter 3). As the time period is small, I would recommend focusing mainly on your notes and brushing up the concepts and formulae in this gap, as the majority of JEE Mains questions are formula-based.

Start with Class 11 topics notes, as you have already covered Class 12 topics while preparing for the Boards. You can also practise one or two sample JEE Mains pattern tests to get back into the rhythm of answering objective questions and your test strategy.

Keep the last 3–4 days before the JEE Mains exam stress-free and make an effort to finish all major revisions and sample practice tests before this period. Reserve the last 3–4 days to revise your strong topics. This will increase your confidence level.

Note: If you have already completed the Class 12 Board exams in the previous year and have no liability of writing the Boards this year, take advantage of the opportunity. Make an elaborate revision plan for the JEE Mains akin to the revision plan of JEE Advance, which will be explained in the next section. Also, practice at least 4 to 5 sample JEE Mains pattern tests in 3 hours' time duration in the last 30 days prior to the JEE Mains exam.

If you have opted for the offline mode, practise the sample test by filling your answers in the OMR sheet for simulating the actual exam conditions. And if you have opted for the online exam, practise the sample tests online.

Consider taking these tests from 9:30 am to 12:30 pm to coincide with the JEE Mains exam timings, so that your mind and body get prepared accordingly.

JEE Mains to Advance

There is a gap of 45–50 days between the JEE Mains (JM) and JEE Advance (JA). After the JM, most of the candidates calculate their marks based on the answer keys given by coaching institutes till the official answer key is announced. Candidates who find their marks on the borderline of qualifying for the JA (based on last year's cutoff marks) often ask, "Should I start preparing for JA, or wait for 20–25 days till the JM result is declared?" Their dilemma is understandable, but what is the harm in starting preparation?

Imagine you discover you have qualified for the JA but have wasted half the preparation time waiting for the results. So start revising as though you have qualified. Start revising from day 1.

Revision Plan

It is better to move in a planned way than by randomly picking topics and revising. In 45 days, it would be very difficult to revisit and solve everything you did in 2 years. However, the *markings* done by you will be your savior in selecting the worthy Qs, ensuring complete coverage of the syllabus in limited time. The revision strategy of categorizing topics into weak/medium/strong is explained in detail in Chapter 4 (in the section *Revision Strategy*).

In the first 30 days, I would recommend that you focus more on practising and solving Qs rather than revising notes. You have already gone through your notes for the JEE Mains preparation. Again, during the last 15 days, shift your focus to the notes.

Notes are generally the most comfortable thing to study, and revising them increases confidence better than anything else for most candidates. Also, notes give a feeling of fulfillment and coverage of the complete syllabus.

If you have gone the extra mile and made some handwritten formulae lists, short notes, or charts, go through them, as they are your most valued assets gained from hard work.

Train Your Body

"In the last month after the board exams I used to wake up at 6 am, have breakfast, study without any break from 9 am to 12 pm, take lunch, and then again study from 2 pm to 5 pm or take a mock test using table and chair, as if sitting for the final JEE exam. I followed this for 15 days before the final JEE day." These are the words of Achin Bansal (AIR-1, JEE-2007). This body training is followed not only by Achin, but by many others who have cracked the JEE.

Why is body training important for the JEE? Our body has a lot of inertia and it performs as it is trained. The JEE has two papers: the first one from 9 am to 12 pm and the second one from 2 pm to 5 pm. On the JEE day, you would undoubtedly want your body to perform at its best during the exam timings. So, it needs to be trained accordingly, and the exam day schedule needs to become an internalized routine of your body.

Simulate the JEE day exactly for at least 15 days before the final day. Wake up at 6 am or 7 am as you will on the JEE day. Have a light breakfast. Sit at a stretch from 9 am to 12 pm as though giving the actual JEE, without taking breaks or snacks or seeing mobile notifications. From 2019, the JEE Advance has been declared to be conducted online. So, if you are taking a test from 9 to 12, it is recommended that you take a mock computer-based test.

Use the timer of the test platform and not your watch, as wristwatches are not allowed in the JEE (it was not allowed in 2017). After the first round, take a healthy lunch before 1:30 pm (but not very heavy, as it may make you doze off). Get ready for the second round.

Study or take a test at a stretch from 2 pm to 5 pm. Analyse what went Chapter 13—The Last Month 125 wrong and improvise the next day. Practise the same schedule for 15 days and your body will be prepared for peak performance on the destined day!!

Some students are accustomed to studying till late in the night and waking up late in the morning. However, for the JEE, as you need to wake up early in the morning and be fully awake and at your best during the exam hours, change your schedule as per the JEE day at least 15 days before the exam. The body needs time to adjust.

Sportsmen train their bodies for their game, soldiers train their bodies for war, astronauts train their bodies for space, and YOU need to train your body for the JEE.

The Day Before Exam

The best thing about a day before the exam is that it is the last day of study for that exam. And when it comes to the JEE Mains or JEE Advance, studying on the day before exam doesn't matter much. What matters is your preparation of two/three years. Just chill on the day before the exam and keep your head cool.

Many students do not study a single word a day before and perform very well. However, it is not a hard and fast rule. If you feel like studying, study. But study the topics you are comfortable with. Skim through your notes or formulae. Do not study topics that you are uncomfortable with.

Assume that those are not going to come. Doing a new Q on the last day is not recommended, as in case you get stuck with it, you may not find anyone to clarify it, which will unnecessarily increase butterflies in your stomach.

Avoid all those things that may increase your anxiety. The list includes talking to people (may be some of your classmates) who always carry a bunch of worries with them and share them generously. Shoo them away. If your exam centre is in a different city, make sure you get there a day before the exam and not later than evening. I would also recommend visiting the exam centre on the day before the exam to gain some familiarity with the location and avoid panic the next day as you hunt for the centre. If you do not feel like going and want to relax, your guardian can visit.

Before going to bed, make sure you have kept everything needed (like your admit card, stationery, etc. and not to forget confidence) for the next day and yippee! you are now free from studies. With this contented feeling, sleep properly at night.

To ensure sound sleep at night before my JEE exam, I made my body a bit tired by doing some light exercise in the evening and took a walk after dinner. But not everything goes as planned sometimes. While I was sleeping, at around midnight, a *baraat* (marriage procession) came below my house out of nowhere with their *band, baja*, and *desi* songs blasting in loud volume. My sleep was disturbed!!! For around half an hour till the procession went away, the music did not let me sleep.

Sleep sometimes can be non-cooperative. Even after that half an hour, I was tossing in the bed for another half an hour, but could not sleep because of irritation and anxiety. So I switched on my DVD player for better songs and started dancing to get my body exhausted.

Fortunately, after 10 minutes of hearing my music player, my mother came to my room. She put my head on her lap, caressed my head, and magically, I went off to sleep. I can never thank her enough for all that she has done for my career. The next morning when I woke up, though the sleep was not as deep as I would have liked it to be, the excitement and anxiety of the exam with its consequent adrenaline rush kept me alert and responsive.

The JEE Day

> ***Have faith in yourself, all power is in you, be conscious and bring it out.*** —Swami Vivekananda

This is the day you will manifest your destiny. Walk to it in confidence and conviction. The days and months and years of hard work, sacrifice, body training, long study hours, test strategies, concepts, tricks and tips to solve the questions will all stand beside you like friends, like allies.

It is your day, the day for which you have worked so hard, the day worth remembering irrespective of the outcome. This day will obviously make you a bit nervous, but that is perfectly fine and helpful too. Have a light breakfast, as you have trained your body. Reach the exam centre timely and be cool.

As you go to your allotted room, it is a good idea to see where the washroom is, so in case you need to use it during the exam, you do not waste time searching for it.

As you begin your exam, if you feel anxious and experience jitters (which happens to many people) during the first few minutes, hold on a minute, smile at the paper, make friends with it, and then crack it with steady hands! Similarly, if in between you get exhausted or your mind becomes blank, take a short break. Close your eyes, take a deep breath. This one-minute break will help you regain mental calm.

Do not panic if the first Q you attempt is very difficult and you are not able to solve it. It might be a trap set by the paper-setters to get you caught at the first ball of the match. Many candidates approach the first Q egoistically and spend a lot more time on it than it is worth. Remember there are other Qs also waiting for you to score. So move on!!

After the first paper is over at 12 pm, you will probably leave the exam hall by 12:15 pm. You will have to go in for the second paper before 2 pm as per the instructions. So, effectively, you will get a break of less than two hours. After the first paper, it is very tempting to find out how it went. It is normal human behaviour. However, it would be wise not to fall prey to this temptation.

Discussing the first paper in the break may give you unnecessary anxiety, which is not going to be of any good for the second paper. Whatever has happened, has happened. Assume it happened for the best. Going to the second paper with the aftermath of the first one in mind is only going to be a distraction during the second paper.

Wisely use the break. Eat something light, refresh yourself, and calm your mind for the second paper. Also, the JEE time is held in peak summers in most of India, and some of the JEE centres might not have an appropriate nearby place to sit and relax during the break. In such cases many candidates use their air-conditioned cars to beat the heat and quietly relax. Get some sleep if you can. Yippee!! You are all set for the second paper.

In the second paper, you are already an experienced person and familiar with the surroundings. Give it your best. You will not even realize how those 3 hours will fly by and when the clock ticks to 5 pm, and your JEE journey comes to an end. You have accomplished the task of writing the highly competitive JEE exam. Congratulations! Party time!!

Scan the QR code to listen to the author on this topic

14

The Art of Writing the Boards with JEE

How to manage the Boards while preparing for the JEE? How to maintain a balance between them? How and when to give time for Board preparation and when to focus on the JEE? These questions start coming to the mind of a JEE aspirant as soon as s/he enters Class 12. Not to forget the BIG question: Is it possible to simultaneously do well in both the Boards and the JEE?

The answer is "Yes." A definite "Yes." In the past, many students of Class 12 have cracked the JEE with a good rank and scored well in the Boards, which means it is very much possible. Let us see how.

First of all, I bring focus to the misplaced perception that most candidates have: the Boards and the JEE are two completely different compartments. Subconsciously, the Boards and the JEE are visualized as shown in Fig. 14.1. But if we look carefully, we will soon find a big overlapping region: physics, chemistry, and maths (PCM) of Class 12.

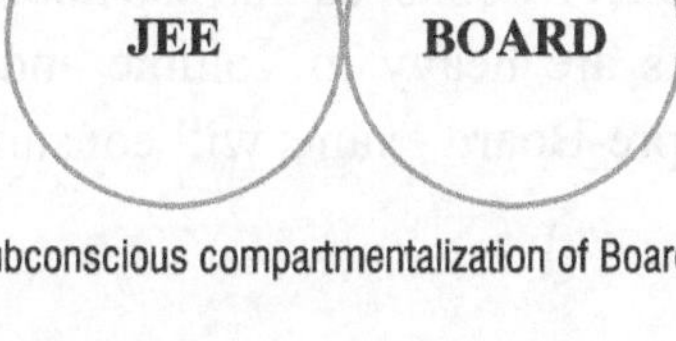

Figure 14.1 Subconscious compartmentalization of Board exam and JEE

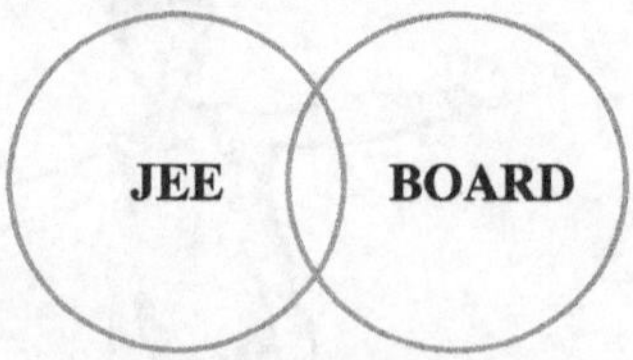

Figure 14.2 Overlapping of the syllabi of the Boards and JEE

PCM are common in the Board exams as well as the JEE (see Fig. 14.2). The syllabus from Class 12 for these subjects is almost the same for both the exams. The difference is that the questions that come in the JEE are more challenging as compared to those in the Boards. The difference is in the complexity and level of challenge, not the basic content. So, preparing for the JEE automatically covers most of the academic content required for the Boards (PCM). Half of the job is over already!!!

The other half The Boards have subjective answers and the JEE is an OMR-based objective exam. Also, the questions in the JEE are harder to crack than those of the Board exams. So, the material for preparation for both is different. For the Boards, NCERTs (or the respective Board-specified books) are the major ingredients, and for the JEE, books having objective problems or coaching material are followed. For a JEE aspirant, generally, most of the time from April to December of Class 12 is consumed in preparing for the Class 12 syllabus of JEE, and thus the material specific to the JEE takes the lion's share of time. In a strategy commonly followed, studying NCERT/Board books and preparing for the Boards is left for the last couple of months before the Board exams, i.e., January to February. In this strategy, the months January and February have to be completely devoted to NCERT books, as this has not been done before. The NCERT books are heavy in volume and language. Also, the practical and pre-Board exams will consume time.

Thereafter comes the month of March and the Board exams. Then one or two weeks after the Boards comes the JEE Mains in April. It has the syllabi from Classes 11 and 12. Alas, for the past three months (Jan, Feb, and March) no time could be spared for brushing up the Class 11 syllabus, and only a little time is left for the JEE Mains. This could be an alarming situation!

The strategy of leaving NCERT/Board books completely for the end has a risk that we might not want to take. So, what is an alternative strategy? You will find it in the next few sections.

When and How to Read NCERT Books

When to read As soon as a chapter of the Class 12 syllabus gets finished in your JEE routine, read that chapter from the NCERT (or Board-specific) book. This will serve a two-fold purpose. First, NCERT books are rich in content. They will act as a supplement of theory and enhance your understanding of concepts from the JEE viewpoint.

Second, you will finish a chapter for the Boards. Regarding solving problems, the Additional Exercises in physics and Miscellaneous Exercises in maths NCERT books have good questions. I would strongly recommend doing these exercises for JEE. If you want some basic practice for the Boards, refer to the general exercises of the NCERT books.

How to read There is an effective reading strategy. Do not dive into reading directly. First, take a glance at questions that came in previous year Board papers. For this, get a book that has chapter-wise segregated previous years' Board questions. You will get an idea of the important profiles (I have talked about previousyear questions in detail in the next section). Now read. While reading, do not forget to mark (explained in Chapter 3). Mark things from two perspectives: Boards and JEE.

There would definitely be some good points that you would like to revise for the JEE. Mark them with a colour highlighter, and mark the things important for the Boards with some other colour.

Advantages There are many advantages of this strategy:

1. In January–February, as you are already done with one reading of the Board books and have the relevant things marked, Board preparation will take lesser time and you can spare time for the Class 11 syllabus of the JEE. This will keep you in touch with the JEE syllabi of both the classes.
2. For JEE, you have a marked-up collection of fine and important points from the Board books, which you can revise in a short time.
3. As you have once gone through the Board books, your confidence level will be high during the crucial months of January–March.

Preparation Specific to the Boards

Along with preparing with books and your notes, do have a look at the blueprint issued by your Board (some of the boards issue it on their website). For each subject, the blueprint has details of the marks and number of questions that each chapter/unit would carry in the exam. This blueprint will give you an idea of the relative importance of chapters and an overall picture for a subject. I would also recommend looking at official sample papers issued by the Board.

Along with this, some of the boards also give the step-wise marking scheme in these sample papers or you may get it in previous years' marking schemes issued by the Board. The step-wise marking scheme will give you an idea of how much is to be written for the questions in each section and which steps will fetch how many marks. An example of a 4-mark question and its marking scheme from the 2016 Mathematics CBSE (Delhi) paper is given below[1]:

Q. If $y = x^x$ prove that $\frac{d^2y}{dx^2} - \frac{1}{y}\left(\frac{dy}{dx}\right)^2 - \frac{y}{x} = 0$

Making scheme

$$y = x^x \Rightarrow \log y = x \cdot \log x \qquad \ldots\frac{1}{2}$$

$$\Rightarrow \quad \frac{1}{y}\frac{dy}{dx} = (1 + \log x) \qquad \ldots 1\frac{1}{2}$$

$$\Rightarrow \quad \frac{1}{y}\frac{d^2y}{dx^2} - \frac{1}{y^2}\left(\frac{dy}{dx}\right)^2 = 1\frac{1}{x} \qquad \ldots 1\frac{1}{2}$$

$$\Rightarrow \quad \frac{d^2y}{dx^2} - \frac{1}{y}\left(\frac{dy}{dx}\right)^2 - \frac{y}{x} = 0 \qquad \ldots\frac{1}{2}$$

As they say, old is gold—remember there is a golden mine in the past papers. Some of the profiles/questions have shown their faces repeatedly in the past papers (though not in exact ways). An example from physics paper (CBSE) is given below:[1]

Explain the principle and working of a cyclotron with the help of a labelled diagram. A cyclotron's oscillator frequency is 10 MHz. What should be the operating magnetic field for accelerating protons? If the radius of its 'dees' is 60 cm, what is the kinetic energy of the proton beam? (CBSE 2005)

Draw a neat and labelled diagram of a cyclotron. State the underlying … (CBSE 2006)

Explain, with the help of a labelled diagram, the principle and construction of a cyclotron… (CBSE 2007)

1 These questions have come in CBSE papers of different zones over the years.

Draw a schematic diagram of a cyclotron. Explain its underlying principle... (CBSE 2008)

Explain the principle and working of a cyclotron with the help of a schematic diagram... (CBSE 2009)

Draw a schematic sketch of a cyclotron. Explain clearly the role ... (CBSE 2013)

Draw a schematic sketch of a cyclotron. Explain, giving the essential details ... (CBSE 2014)

A cyclotron's oscillator frequency is 10 MHz. What should be the operating magnetic field for accelerating protons? If the radius of its 'dees' is 60 cm, calculate the kinetic energy (in MeV) of the proton. (CBSE 2015)

State the principle of a cyclotron. Show that the time period of revolution of particles... (CBSE 2016)

There are many such examples. It appears there are some favourite questions of your Board and it would be a shame to not recognize them and miss working on them. Go through the previous-year papers (3–4 years or more as time permits) and identify such profiles/questions. At the same time, get an idea of the difficulty level.

If you are studying under the CBSE, I would additionally recommend going through previous 2–3 year papers of all the zones including your zone.[2] Questions repeat inter-zone also. In the case of the State Board, you might not have the option of reviewing papers of other zones, but you could go through the paper that came in the supplementary exam (it does not mean that you will be writing the supplementary exam; the paper helps you to know the question profile preferences of your Board).

2 CBSE webpage for previous year papers - http://www.cbse.nic.in/student.htm. It has many useful things including the marking schemes for previous year papers.

In the Boards, in addition to PCM, there are two (or more) other subjects also. One is English for most of the Boards. English will demand some of your precious time. Give it in small chunks. While preparing for the JEE, when you need to take a break from studying PCM, read a story from your English book. Or write something in English. These small investments will not hamper the JEE preparation and will, instead, pay you well later. The other subject(s) can be any of the following: physical education, Hindi, computers, or other subjects, depending on the Board. In my case, I opted for physical education as the fifth subject. Generally, students preparing for the JEE and medical entrance exams opt for physical education, as it is a relatively lighter optional subject than others. The guidance of teachers of respective subjects can be beneficial and time saving while choosing optional subjects. Do not hesitate in taking help from them.

The pre-Board exams conducted by your school (nearly a month before the Board exams) can help you and serve as rehearsals for the Boards. They will give you a practice of writing subjective answers, which might be lost in preparing for the JEE. These papers will also help you gauge your Board preparation level.

When your answer sheet gets evaluated, look for each area where you have lost marks and work on their rectification. If you need more practice, give some of the previous year Board papers within the time limit set for Board exams. Get these answer sheets strictly evaluated by your teachers, take detailed feedback, and work on them. You are ready for the Boards!

Writing the Board Exams

The Board exam is a subjective exam in which you have to write subjective answers, which the paper-checker has to review in detail. Given the large number of people who take the Board exams each year, each checker is allotted several answer copies to be reviewed within a short period of time. We should also not forget that the checkers are humans and the copies are evaluated when the summer is at its peak in most parts of our country. In such demanding conditions, if going through your answer copy gave further stress to the checker, then God bless you!!

Your answer copy should be checker-friendly. The things that get noticed in the first instance are the formatting, diagrams, underlined words, and boxed items. The following tips will help you present your answers in a better way:

1 Make diagrams wherever appropriate; it leaves good impression on the checker's mind.
2 In case of a numerical, put your answer in a box to highlight it and write *Ans* beside the box. (Do it only when you are confident of the answer!)
3 In your answers, underline the important words/phrases. Do this only after you have answered all the questions and are left with some time.
4 If reading your handwriting has been a big pain for people (like it was the case with my handwriting before the Class 10 Boards), try improving it at least a bit before the Boards. It can be done. For instance, try giving more spaces between words and see if it improves legibility.

In the Board exams, there are questions carrying different marks and the paper generally has questions in the sequence of increasing marks, i.e., 1-mark questions at the beginning and 5- or 6-mark questions at the end.

Regarding the order of attempt of questions, I would recommend starting with the maximum-mark Qs. In the beginning, when we are full of energy and time is available, we have the tendency to write more in each question.

So, if we start with the 1-mark Qs (which generally have very short answers) and add extra lines than required, we cannot earn more than 1 mark per question. It will mean under-utilization of time and energy. Employ your initial energy for Qs carrying the maximum marks.

Second, in case you have little time left before the bell rings, it will be difficult to complete 5- or 6-mark Qs as compared to 1- or 2-mark Qs. Try these strategies first with sample papers, and if they work well for you, use them in the final exams.

The Boards can be managed efficiently along with the JEE and you can do great in both. Keep your spirits high and prepare smartly. *Be a test warrior. All the very best for the Boards and the JEE!*

Scan the QR code to listen to the author on this topic

15

Life @ IIT

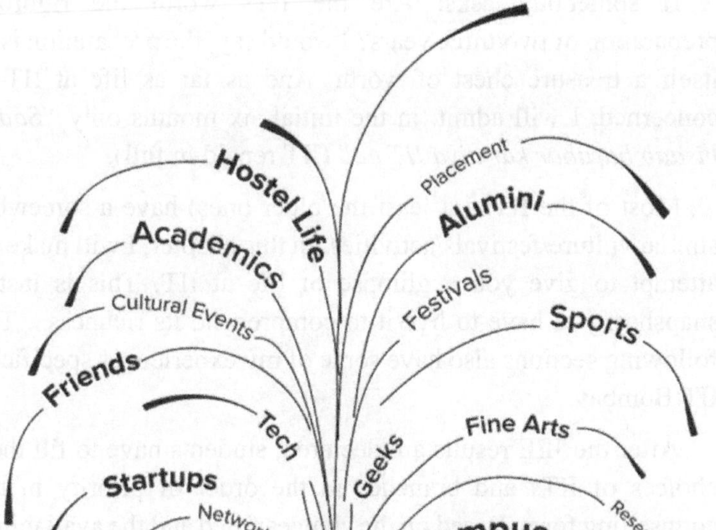

For two years during the JEE preparation you lock yourself up in a room with PCM books and throw away the keys, and you will most probably make it into one of the IITs. At an IIT also, you lock your room and throw away the keys, but this time you are outside the room. There are so many things/activities to explore at IITs that the four-year duration appears barely enough to spend at such an enchanting place.

If somebody asks: Are the IITs worth the rigorous preparation of two/three years? I would say, the preparation is in itself a treasure chest of worth. And as far as life at IIT is concerned, I will admit, in the initial six months only *"Saara hisaab barabar kar diya IIT ne"* (IIT repaid in full).

Most of the IITs (at least the older ones) have a somewhat similar culture/festivals/activities. In this chapter, I will make an attempt to give you a glimpse of life at IIT. This is just a snapshot; you have to live it to comprehend its richness... The following sections also have some of my experiences specific to IIT Bombay.

After the JEE results are declared, students have to fill their choices of IITs and branches in the order of priority in the counselling form. Based on the choices filled and the availability of seats (which get filled based on priorities as per the JEE rank), candidates are allotted the IIT and the branch. New session at IITs commence from around the third week of July.

At IIT Bombay (IITB), the rain gods will inevitably be there to greet students to the lush green campus widespread across 550 acres, with a lake inside the campus. The campus is serene, and people fall in love with it. Students are registered and hostels are allotted. The IITs have their own language known as IIT lingo. For instance, first-year students are known as 'Freshie' in IIT-B lingo.

When students get into IITs, many parents presume that their ward's career is all set now, and students consider IIT as the place meant to reap the fruits of their hard work done during JEE preparation.

No doubt, IIT is the much awaited reward of the hard work, but it is not the end of it. It is a new beginning towards a future career. It is the land of opportunities not to be missed.

Talking about academics, four years at IIT would have 8 semesters (10 semesters in the five-year dual-degree program) with a minimum number of courses/credits to finish in each semester. The academic program is robust and the facilities are the best you can get in the country, be it labs or lecture halls (air-conditioned) or auditoriums. You will have every opportunity and resource to pursue your academic interest and research under the guidance of outstanding and internationally reputed faculty. The huge central library at IITB is no less than a wonder, with around 5 lakh books, journals, theses, and what not, covering nearly every genre, be it literature, history, fiction, biographies, engineering, management, etc. There is no lack of avenues—you can do as much as you want or, I must say, as much as you possibly can.

After each even semester, there is a break of around three months in which students explore different things such as taking projects at IIT, going to mountaineering camps, preparing for festivals at IIT, or enjoying the holidays at home. After the sixth semester, the students go for an internship. Some of the students opt for getting exposure to the respective industries of their choice, some take up research projects at IIT, and some opt for foreign universities. The internship program is like a mini placement program supported by the IITs, where companies and universities come to campus and select students. Exposure at internships helps students in choosing their future career path. After the seventh semester, companies are invited to the IIT campus for the muchawaited job placements. Students are interviewed and some of them are offered such fat pay packages that make newspaper headlines!

While academics and research at the IITs are remarkable, the avenues and events for all-round personality development make them top-class. Be it sports, cultural, technical, social service, entrepreneurship, gaming, or even fashion, the IITs have them all. The JEE preparation days might have all been spent with books, but this will not be the case at the IITs. IITians enthusiastically pursue their extra-curricular interests and explore new fields. For instance, at IITB, there is a super-active club for nearly everything, be it dance, dramatics, theatre, gaming, coding, rural activities, astronomy, entrepreneurship, cooking, music, photography, fine arts, debating, and a lot more. If there is none matching your interest, you can create a new one!

These clubs conduct many exciting events. As far as I can remember, we had some or the other event happening at IITB each day, leaving only the exam days. There were times when we had to skip some of the events as some other event was happening at the same time. Sometimes, we would get to know of the events via our newsletters. Yes, IITB has its own newsletters written and published by the writing-enthusiast students. It was after the first semester at IITB that we realized that it was not at all possible for anyone to take part in everything, and we learnt to prioritize which events to attend and which ones to skip.

We learnt time management by waking up as late as at 8:20 am and then brushing teeth and stuffing breakfast on the way to class at 8:30 am (as we were busy preparing till late the previous night for some festival/event the next day). And to add to the fun and food, our hostel canteen would remain open till 4 am at night (or morning). In the next few sections I will discuss features specific to IITB. Other IITs also have similar festivals/events but known by different names.

Cult IIT Bombay cultural[1] (known as *cult* in IITB lingo) has ten genres—music, dance, photography and fine arts, dramatics, speaking arts, literary arts, film & media, classical & folk arts, design, and lifestyle. Various events like the PAF (performing

1 https://gymkhana.iitb.ac.in/~cultural/

arts festival), Freshiezza (festival for first years to showcase their talents), Goonj (students' band singing competition), AIDS (Annual InSync Dance Show) etc. are conducted by these clubs at hostel and institute level.

IITB is host to Asia's biggest college cultural festival *Mood Indigo* (known as MoodI or MI[1]), which had a footfall of more than 1,39,000 students from 1700+ colleges pan India in the year 2016. The four-day fest, which happens in December, has witnessed more than 230 events in the year 2016, the most popular event being *Pronites*, where renowned artists and bands of the likes of Shankar-Ehsaan-Loy, Sonu Nigam, Zakir Hussain, Porcupine Tree, and Borgeous, to name a few, have performed in different years.

Tech *Hum science ki taraf se hai*. IIT is Indian Institute of Technology, so how can IITs be behind in technology (Tech). IITB hosts *Techfest*—Asia's largest science and technology festival[2], which had a footfall of more than 1,65,000 people from over 2500 Indian colleges and 500 overseas in the year 2016–17. It is a threeday long festival having exciting and challenging competitions such as Robowars (fighting robots), Aviator (design and fly your own aircraft), Coding Arena (who is the best coder), and many more to find how big a geek you are!

Many eminent personalities including many Nobel Laureates from around the world are invited to give lectures. In the evenings, Technoholix shows such as laser shows, tron dance, virtual 3D holographic displays, etc. thrill the crowd with the wonders of technology.

To nurture the spirit of innovation, the National Innovation Club has been founded in IITB to encourage students to step up

1 https://moodi.org/#/

2 http://www.techfest.org/

and find innovative solutions for the socio-technological challenges our country is facing. This group is a part of the social service cell named NSS (National Service Scheme) where many projects are ongoing for the betterment of the society.

IIT Bombay has also left its mark in satellite and space technology through its first student satellite 'Pratham' launched by ISRO in 2016. Further, IITB has a racing team that builds actual race cars. This team has participated in international student competitions such as Formula Student held in the UK and won it too!!! Would you not want to be a part of a satellite mission or racing event some day?

Entrepreneurship Ola Cabs, Flipkart, Myntra, Toppr, InMobi, Zomato, etc.—Yes, you guessed it right. These are all startups founded by IITians. If you have the entrepreneurial streak, the IITs will not disappoint you. There is a dedicated cell for promoting entrepreneurship (at IIT-B it is known as the E-Cell[1]), which conducts events and workshops to expose students to the startup world, helps them connect to successful entrepreneurs, and learns under their mentorship. IITB has this society named SINE[2] (Society for Innovation and Entrepreneurship) which incubates early-stage technology startups and assists them with resources, networking, and mentorship.

Sports Life at IIT is unimaginable without sports. With world-class courts, playgrounds, and other sports infrastructure, the IITs are home to a wide range of both indoor and outdoor sports. Be it lawn tennis, cricket, football, hockey, squash, table tennis, swimming, Kho-Kho, billiards, athletics, chess, and even fooz-ball, the IITs have them all and that too with a dedicated coach for each of the sports[3]. Apart from the central gymkhana,

1 https://www.ecell.in
2 http://www.sineiitb.org/about.html
3 https://gymkhana.iitb.ac.in/~sports/

which has all the sports amenities and grounds, there are separate grounds, table tennis tables, billiards tables, etc. in each of the hostels! Playing football the whole night in the muddy hostel grounds under the shower of rains is just unforgettable. Many tournaments are conducted at hostel and inter-hostel level, giving ample opportunity to every sports enthusiast to participate. Not to forget the Inter-IIT tournament, where the best sportsmen from each IIT in different sports meet in one of the IITs, compete, and win medals for their respective colleges. The awesome part of the IITs is that all these facilities are open to alumni even after they pass out.

The most important aspect of all these extra-curricular activities is they are managed by the students. Students hold what is called the PoR (position of responsibility) for each area. For example, at hostel level, there are PoRs such as Mess Secretary (Secy), who takes care of the mess affairs (food), and Sports Secy, who plans and organizes sports tournaments in the hostel and represents her/his hostel at the institute. Similarly, dramatics, music, computers, and even the library have a secretary to take care of its respective portfolio at hostel level.

At institute level, there are other portfolios such as sports, academic affairs, hostel affairs, and cultural affairs, which are managed by a Secretary who you can think of as a CEO of that portfolio. Then there are sub-posts under each portfolio such as a Secretary for Cricket, Tennis, etc. at the institute level, who plans and organizes events at institute level, coordinating with the secys of the hostels. There is a whole council of students who are elected by voting (yes, elections happen at IITs also), and who take responsibility and spare no effort to make their portfolios a grand success. A similar structured council exists for events such as Mood Indigo, Techfest, E-Cell, etc., who take care of every nitty gritty of the event, ranging from contacting artists, getting sponsorships, designing, marketing, and what not. For Techfest

2016–17, a team of 800 students worked day and night to make Techfest a splendid spectacle.

These portfolios impart invaluable managerial experience, exposure, and skills of planning, execution, and working in a team, and other important leadership qualities. Who would have known that once a hostel Mess Secretary in IITB, Manohar Parrikar would be the Defence Minister of India, or an organizer of Mood Indigo, Nandan Nilekani, would lead the massive UID (unique identification) project.

Four years at a Brand IIT college with the smartest of peer groups, strong academics from outstanding faculty, variety of extracurriculum activities, and fun-filled hostel life develops one's personality as a whole and provide opportunities to explore different arenas. Based on these learnings, experiences, and interests developed at IIT, students choose different career paths after college.

Some opt for placements in core departmental companies; some go for financial sector, marketing, FMCG, etc. Some of the students take up research and higher studies (MTech, PhD, MS, or MBA) and some prepare for administrative services. Some fearless ones start their own ventures. A few also take unconventional paths like Nitin Gupta (known as Rivaldo), a Chemical Engineer from IITB but much of a dramatics fanatic, is now a stand-up comedian by profession. Whatever career path the students decide to take, it is very unlikely that they will not found alumni of the same area to take guidance from. Strong alumni network is no less than a boon for both the IITs and its students, who make the IITs what they are.

If by now you have started to believe that IITs are the only gateways for a bright future and success, then you have got it wrong dude! Think about it. IITs are just one of the options. Satya Nadella, after graduating from a non-IIT college, became the CEO of Microsoft. Paytm founder Vijay Shekhar Sharma did his graduation from Delhi College of Engineering.

There are many such people out there who made it without IITs. So what/who is the reason of success? I believe it is the qualities of people that makes them what they are, and in today's scenario there is no lack of opportunities for a determined person.

No doubt, the IITs provide a conducive environment and the required exposure to gain the needed qualities, but again at the end it is the individual that has to earn them through his/her hard work.

After getting out of IIT, the brand IIT may give an individual a head start, but after that, it is only the individual's capabilities and qualities that take her/him to glory. People like Mark Zuckerberg, Bill Gates, and Steve Jobs did not even wait for their college graduation to complete, and have achieved what most graduates cannot even dream of.

So when it is a matter of qualities, why not start inculcating the traits of hard work, perseverance, sincerity, planning, and strategy right from the JEE preparation phase itself. These are the formative years of your personality, and the invaluable traits earned during these years cannot be taken away, whatever may be the result. I would say: now that you have chosen the path of the JEE, give it your best, be strategic, planned, hardworking, and keep earning qualities. My best wishes. All the best!!!

Scan the QR code to listen to the author on this topic

